NULLIFICATION AND SECESSION IN THE UNITED STATES

A HISTORY OF THE SIX ATTEMPTS DURING THE FIRST CENTURY OF THE REPUBLIC

BY

EDWARD PAYSON POWELL

THE LAWBOOK EXCHANGE, LTD.
Clark, New Jersey

ISBN 978-1-58477-132-6

Lawbook Exchange edition 2002, 2017

The quality of this reprint is equivalent to the quality of the original work.

THE LAWBOOK EXCHANGE, LTD.
33 Terminal Avenue
Clark, New Jersey 07066-1321

*Please see our website for a selection of our other publications
and fine facsimile reprints of classic works of legal history:*
www.lawbookexchange.com

Library of Congress Cataloging-in-Publication Data

Powell, Edward Payson, 1833-1915.
    Nullification and secession in the United States : a history of the six
    attempts during the first century of the Republic / by Edward Payson
    Powell.
        p. cm.
    Originally published: New York: G.P. Putnam's Sons, 1897.
    Includes index.
    ISBN 1-58477-132-1 (cloth: alk. paper)
        1. Secession. 2. Nullification. 3. Constitutional history—United
    States. I. Title.

    KF4613 .P69 2001
    973.5—dc21   00-067013

Printed in the United States of America on acid-free paper

NULLIFICATION AND SECESSION IN THE UNITED STATES

A HISTORY OF THE SIX ATTEMPTS DURING THE FIRST CENTURY OF THE REPUBLIC

BY

EDWARD PAYSON POWELL

" An Indissoluble Union under one Federal Head."
GEORGE WASHINGTON, 1788.

"A Federal Republic in which States Form the Central Principle."
GEORGE CLINTON.

" Freedom of Religion ; Freedom of the Press ;
Freedom of Commerce."
THOMAS JEFFERSON.

G. P. PUTNAM'S SONS

NEW YORK   LONDON
27 WEST TWENTY-THIRD STREET   24 BEDFORD STREET, STRAND
The Knickerbocker Press
1897

The Knickerbocker Press, New York

To the South as well as the North, to the West as well as the East, with a spirit of equal justice to all, this book is dedicated—our common land—our indissoluble Union.

"With malice toward none; with charity for all."

ABRAHAM LINCOLN.

PREFACE

THIS book is undeniably written for a purpose. It is my desire to state facts as viewed from a strictly national, rather than from a sectional or partisan standpoint. But it is not without the sphere of legitimate history to aid by such a statement of facts, in creating a more generous national sentiment; and a conviction on the part of all sections that political righteousness has not been the exclusive property of any one part of the United States. It is time to deal justly by the South; and recognize its full share in the better part of nation building; while at the same time we do not overlook the diverse obligations that naturally fell upon the complementary sections. It is easy for either North or South to perceive the blunders in action and defects of character of the opposite section; it is difficult to generously measure each other's achievements; and to help atone for each other's errors. In writing a history of attempts at nullification and secession, I shall not forget that they are expressions of that intense individualism which was the most potent factor in creating our Republic.

E. P. P.

CONTENTS.

CHAPTER I.

CHAPTER II.

CHAPTER IV.

The Union vastly enlarged by the Louisiana Purchase—Burr in desperation turns to the Southwest—
He forms alliance with General Wilkinson ; is be-
friended by Jackson—He aims at a great Mississippi
Valley Confederation—The people true to the Union—
Wilkinson deserts Burr—Burr plots in all directions ;
starts to New Orleans with a flotilla ; is arrested—
Burr's trial at Richmond a fiasco—Burr goes to England ; is ordered out of the country ; goes to France ;
is kept under surveillance—Released, he hurries to
America—His later history one of wretchedness—Is
buried as a pauper.

APPENDIX TO CHAPTER IV.—(1) Burr's Valedictory to the United States Senate. (2) President Jefferson's Message on the Burr Conspiracy. (3) Burr
at his Trial. (4) Testimony of William Eaton. (5)
Burr and Seward.

CHAPTER V.

Interference with Neutral Rights by England—The
Berlin and Milan Decrees of Napoleon—Congress
declares war against Great Britain—New England
protests—Efforts to thwart our enlistments—Disaster
on land, but success on the ocean—Treaty negotiations—England's demands ;—New England Federals
urge their acceptance—The Hartford Convention—
Resolutions looking towards separation—The triumph
of American diplomacy at Ghent, and of American
arms at New Orleans.

APPENDIX TO CHAPTER V.—(1) Report of the
Hartford Convention. (2) Resolutions of the Hartford Convention. (3) Action of Towns.

CHAPTER VI.

CHAPTER VII.

CHAPTER VIII.

NULLIFICATION AND SECESSION IN THE UNITED STATES

CHAPTER I

INTRODUCTORY. DIFFICULTIES INVOLVED IN BUILD-
ING THE REPUBLIC

IN the unwritten Constitution of England, it has
always been a fundamental principle that there
shall be no taxation without representation. Charles
II. said in 1676 : "Taxes ought not to be laid on the
inhabitants and proprietors of the colony (Virginia),
but by the common consent of the General Assembly."

At the close of the war with France, involving their
colonial possessions on this continent, England found
herself in debt. Parliament, in conjunction with the
ministers of George III., passed an act to compel the
colonies to aid her in paying this debt. The colonists
denied the right and legality of any such enactment,
while they were not represented in Parliament. It
was at least a new departure in English legislation.
In 1761, Otis published in Boston *The Rights of the
Colonies Asserted aud Proved*, a book that became a

political Bible. Virginia, in the spring of 1765, by her House of Burgesses, resolved that " the General Assembly of this Colony has the sole right and power to lay taxes and impositions on the inhabitants of this colony." Patrick Henry, a young lawyer, by his defense of this resolution, became known as the most wonderful orator in the world. Lord Chatham, in a speech in Parliament of astounding courage and eloquence, replied to the Ministry that they would sooner or later be compelled to retract ; and that for his part he rejoiced that the colonies resisted.

Massachusetts invited a General Congress of the colonies to meet in New York, October 7, 1765. Nine colonies which were there represented, Massachusetts, Rhode Island, Connecticut, Pennsylvania, Maryland, South Carolina, Delaware, New Jersey, New York, and New Hampshire, agreed to abide by the action of their delegates. Brigadier-General Ruggles was elected President ; but Adams declared that " he ran away " to avoid responsibility. This Congress, the first in American history, issued a protest against offensive taxation, and dissolved. The Stamp Act was soon rescinded by Parliament as a consequence.

In 1768, English troops were sent over to garrison Boston. A conflict in the streets took place in 1770, and Samuel Adams led in a successful demand to have the troops retired to a castle and barracks. In 1774, continued aggressions of the parent country led to Committees of Correspondence between the colonies ; the avowed object being the creation of a General Congress, to meet annually, to deliberate on their common interests.

The first session of this Congress or assemblage of delegates of the colonies took place in September of the

same year. A spontaneous movement, it continued in
session without dissolving for fifteen years. Legislative, judicial, and executive in its functions, without
certified warrant, it declared the independence of the
colonies and won it. It laid taxes, it conducted war,
it negotiated treaties ; its only credentials of authority
being the exigencies of the times. The history of legislation before or since shows nothing more admirable
for patriotism, or more statesmanlike in conception,
than the documents that emanated from these representatives of the people. The Bill of Rights ; a List
of Grievances ; The Association ; an Address to the
People of Great Britain ; and a Memorial to the Inhabitants of the British American Colonies are papers
unequalled in the political history of the world, except
possibly by rare documents appearing at great intervals.

John Adams, answering the question, Did every
member of Congress on the 4th of July, 1776, in fact
cordially approve of the Declaration of Independence,
replied : " Majorities were constantly against it. For
many days the majority depended on Mr. Hewes of
North Carolina. While a member one day was reading documents to prove that public opinion was in
favor of the measure, Mr. Hewes suddenly started upright, and lifting up both hands to Heaven as if in a
trance, cried out : ' It is done ; and I will abide by it.'
I would give more for a perfect painting of the terror
and horror upon the faces of the old majority at that
moment than for the best piece of Raphael." Jefferson gives a synopsis of the arguments for and against a
declaration. The delegates from Virginia moved it on
June 7th. The debate showed that " New York, New
Jersey, Pennsylvania, Delaware, Maryland, and South
Carolina were not matured for falling from the parent

stem ; but as they were fast advancing to that state, it
was thought prudent to wait a while for them."
Meanwhile, John Adams, Dr. Franklin, Roger Sher-
man, Robert R. Livingston, and Thomas Jefferson
were appointed a committee to draft a Declaration to
be considered on July 1st. On that day the delegates
from South Carolina and Pennsylvania voted against
it ; and those from New York, obeying old instructions, did the same, contrary to their private convictions, and to the later instructions of their State. The
States rapidly changed those delegates who were opposed ; and unanimity was reached July 2d. The draft
of a Declaration, submitted by the committee, after
striking out a clause reprobating the slave trade, was
adopted on the 4th ; and on the 2d of August was
signed by all the members present except John Dickinson.

Acting for the States as States, Articles of Confederation were drawn up in July, 1776. But it was 1781
before they were formally adopted by all of the States.
By that time their insufficiency was demonstrated.
Only by consent of nine States could the United States
coin money, borrow or use money, engage in war, or
negotiate a treaty. When Congress acted it had only
the good-will of the people to fall back on. It had no
power to compel the enforcement of a law. Randolph
called it " a government of supplications." It could
neither regulate commerce between the States, nor
affect foreign commerce and make it uniform. While
confederated, the States could and did lay impost taxes
against each other. New York had a tariff on New
Jersey garden stuff and Connecticut firewood. These
States retaliated with similar measures. The smaller
States were equal every way in power to the larger

ones ; and in vital matters, Delaware, Rhode Island, Connecticut, New Jersey, and New Hampshire could negative the action of New York, Massachusetts, Virginia, Georgia, Pennsylvania, and the Carolinas. There was no direct representation of the people. In 1783, an attempt was made by Congress to secure power by impost tax to pay the debt incurred by the Revolution. It was 1786 before the States had all acted ; and then New York had nullified the plan. Resort was had to created or fiat money. Debts were paid in paper. But credit was not given ; and no one trusted his neighbor. The end was insurrection and anarchy. The States began to see that it must be a closer union, or no union at all.

Jefferson, speaking of the Confederacy, says : " The fundamental defect was that Congress was not authorized to act immediately on the people, and by its own officers. Their power was only requisitory, and these requisitions were addressed to the several legislatures, to be by them carried into execution without other coercion than the moral principle of duty. This allowed a negative to every legislature, on every measure of Congress ;—a negative so frequently exercised as to benumb the action of the general government. The want, too, of a separation of the legislative, executive, and judicial functions worked disastrously in practice." Washington, as commander-in-chief of the armies, had especially felt the evils of this loose system ; for the appropriations of Congress and levies of troops had been responded to by the States as they individually pleased. It had become with him a ripened conviction that there was no choice but between anarchy and a stronger government. He said in a letter to Jay : " I do not conceive we can long exist as a nation without having

lodged somewhere a power which will pervade the
whole Union, in as energetic a manner as the authority
of the State governments extends over the several
States." In a circular letter to the governors, he declared the first essential to the very existence of the
United States as an independent power to be, "An
indissoluble Union of the States, under one federal
head." Clinton, as Governor of New York, in his
Message of 1780, laid our embarrassments "chiefly to
a defect of power in those who ought to exercise supreme jurisdiction."

An attempt was made to amend the Articles ; but
Rhode Island alone was able to prevent it. Virginia
led the way finally to a change ; while Massachusetts
and New York were strongest in opposition. Webster,
in 1838, said : "The Commonwealth of Virginia is entitled to the honor of commencing the work of establishing the Constitution. There is not a brighter jewel in
the coronet that adorns her brow." Washington was
the soul of the new movement ; Madison its brains.
Bancroft, speaking of the Confederacy, says, "But
with all its faults it contained the elements for the evolution of a more perfect Union." This was demonstrated in the Convention of 1787 ; for the ultimate
Constitution was in fact but a ripening and perfecting
of the Articles of Confederation ; as they had previously
been born of the Articles of Association.

In the Constitutional Convention of 1787, Washington unquestionably had in view a Union that did not
allow any component State after entering it to withdraw or secede. Hamilton's ideal was extreme. He
would obliterate State sovereignty, if not State lines,
fusing the whole population in one unit, with a single
supreme head. Preliminary to such a step he would

have the State governors to be appointees of the United States government. Clinton would, on the contrary, exalt the State, and give the general government only supervisory and very limited control. " I desire," he said, " a federal republic in which the States shall form the creative principle. Every State must be equal and equally represented. Its representatives must look solely to it for their support, and for their instruction. They must collectively vote in obedience to its will and be separately subject to its recall." Jefferson, quite apart from these, laid most emphasis on individual liberty. Being in Paris, he had no hand in framing the new Constitution, and while heartily wishing its adoption, he desired, in addition to its specifications, a Bill of Rights that should stipulate, " Freedom of religion, freedom of the press, freedom of commerce, no suspension of habeas corpus, and no standing army." He also objected to the eligibility of the President to re-election.

Besides the eleven Articles of Hamilton and the propositions of Pinckney there were two plans laid before the Convention : that of Virginia, which was national ; that of New Jersey, which was federative. To compromise and harmonize these two plans required all the self-restraint, tact, and patience of the Convention. The idea of popular representation was finally injected into the old scheme, which had been purely an affair of States. Those who insisted on an equality among the States were placated by creating a Senate which should express the federated principle. This was the first compromise in creating the Republic. At no time was dissolution more formidable than when the smaller States refused to assist in forming a Union which denied them equality with the larger.

But this was not the only compromise required and granted. Two more were demanded by the existence of slavery; one of these, in the apportionment of representatives, permitted a count of three-fifths of negro slaves; the other forbade the abolition of the slave trade before 1808. The debate that ended in these concessions was violent; and it is notable that the strongest opposition was made, not by New England, but by Virginia. George Mason said : " Slavery discourages arts and manufactures. The poor despise labor when performed by slaves. Slaves prevent the immigration of whites who really enrich a country. They produce the most pernicious effect on manners. Every master of slaves is born a petty tyrant." Gouverneur Morris of New York responded : " Slavery is a nefarious institution. It is a curse of Heaven on the States where it prevails. Travel through the whole continent, and you behold the prospect continually varying with the appearance and disappearance of slavery. Upon what principle is it that these slaves shall be computed in the representation ? Are they men ? Then make them citizens, and let them vote. Are they property ? Why, then, is no other property included ? " He would " sooner submit to a tax for paying for all the negroes in the United States than saddle posterity with such a Constitution." In fact, the Constitution never would have been adopted had not New Englanders come to the rescue. Ellsworth of Connecticut retorted that, " As he had never owned a slave he could not judge of the effect of slavery on character ; but he held it would be unjust to South Carolina, in whose rice swamps negroes died so fast, should they be prevented from importing more." Rutledge of South Carolina added : " Religion and

humanity have nothing to do with this question. Interest alone is the governing principle with nations."
He hinted that New England would find the slave trade for her interest, as the negroes would be imported in her vessels. At any rate the South would form a Union on no other terms. Ellsworth wound up the debate with, " Let every State import what it pleases. The morality or wisdom of slavery are considerations belonging to the States themselves." So at the very outset South Carolina and New England were of affiliated temper ; destined always after to be most cordial friends, or most bitter foes.

The Constitution as finally adopted by the Convention was not held by any one to be ideally perfect ; but probably the best that could be secured. Washington, standing with pen in hand to sign it, said : " Should the States reject this excellent Constitution, the probability is that opportunity will never be offered to cancel another in peace ; the next will be drawn in blood." The only parties that entered into the compact of 1788 were the people organized as States. As " free and independent " they had created the Confederacy, and won united independence ; they now yielded none of this independence in forming " a more perfect Union," except that which was specifically written down in the bond. The Constitution carefully enumerated what the States granted to a central government.

In this new and more perfect Union, the people were directly as well as indirectly represented. They elected the executive, and one branch of the legislative department ; while the Senate still stood for the Commonwealth of States. The judicial department, quite apart from popular election, was to be the creation of the President and the Senate ; while the judges thus con-

stituted were out of reach of the people ; except by
means of impeachment, brought about by co-operation
of both Senate and House of Representatives. The
construction of this department of government, especially the tenure for life, showed a lingering timidity
on the part of the Convention to cut loose entirely from
aristocratic methods, and trust absolutely to the com-
mon-sense and self-restraint of the people. Jefferson
was strenuously opposed to the appointment of judges
beyond a limited term. As late as 1822, he was still
of the same opinion. He wrote : " We have erred in
this point, by copying England, where certainly it is a
good thing to have the judges independent of the king.
That there should be public functionaries independent
of the nation, whatever may be their demerit, is a solecism in a republic."

When the Constitution was finally offered for adoption, it was accepted in nearly all cases with qualifica-
tions or recommendations of amendments. There were
not a few who doubted the policy of endeavoring to
form a complete Union of colonies that had such diverse
social customs, religious opinions, and commercial
interests. It was not believed possible that any system
of frictionless co-operation could be devised. Others
preferred at once to arrange for three groups of States :
the Northern, consisting of New England, with New
York, and possibly New Jersey ; the Central, consisting of Pennsylvania, with Delaware and Maryland ;
and the Southern, which should include from Virginia
to Georgia. At least, said a large party, let us not
undertake too much. What has New England in harmony with Georgia or even Virginia ? Our corn and
our agriculture as well as our simplest social conditions
are wholly unlike. The North and the South can

never work in harness. John Adams noted that not only the common people, but the gentlemen of the South differed sharply from those of the North. " I dread," he added, " the consequences of this dissimilitude. Without the utmost caution and forbearance it will prove fatal." Jefferson, after the Louisiana purchase, doubted whether all that could be hoped for was not to create a solidarity in government so far as to include the Atlantic States.

State rights and sectional sentiment were peculiarly strong in New England. It was a people of intense convictions. The anti-federal sentiment was overwhelming. The *Confederate* government had been held to be something of a foreign sort. To overcome this feeling and create a Union spirit was not easy. The Shays Rebellion undoubtedly did much to create a fear that the existing government was too frail. But there was by no means unanimity of sentiment when it was proposed that delegates be sent to a Constitutional Convention. Let us go back to a monarchy, said some ; " Monarchy and order." The long-continued clerical control of New England had fostered a faith in government by classes. " The people " were not trusted.

There really was much excuse for those who protested against a single republic ; for it was a month's tour to get from New Orleans to Boston. Steam had not yet come to bind the sections together. Each locality developing alone, without close association with others, evolved such individual habits and customs as did not adjust themselves easily to neighbors. But King and Gerry soon were converted in Massachusetts, and brought over that State. The federalists did mighty work in New York, and won the State from

apparently hopeless antagonism to favor the new
Union. New Jersey accepted the Constitution unanimously, chiefly as a safeguard against her neighbor's
unjust taxes. But New York voted ratification on the
declared premise that "the powers of government may
be reassumed by the people whensoever it shall become
necessary to their happiness." Other States used similar language. It was clearly understood that those
who put the government together could take it down
again. Rhode Island held off until 1790, owing to her
peculiarly strong sense of individual freedom. It alone
of the States was born of absolute toleration. It feared
alliances, and had good reason for hesitation and timidity concerning new measures.

The Constitution had to combat local jealousies in
every direction. In the Convention, Pennsylvania had
refused to be on an equality with little New Jersey.
Bedford, of Delaware, had declared that that State
would form a foreign alliance rather than enter a Union
in which it should be at a disadvantage with larger
States. In 1782, Pennsylvania had threatened to break
out of the Confederacy and use its taxes for its individual ends. Rhode Island was continually threatening to start off alone. Monroe argued that force might
be used to prevent Pennsylvania from seceding. Jefferson wrote in 1786 : "I fear that the people of Kentucky think of separating, not only from Virginia, but
also from the Confederacy. I should think this a most
calamitous event. Our Confederacy must be viewed
as the nest from which all America is to be peopled."
There was, without doubt, a strong sentiment in the
Southwest in favor of a separate nationality.

Laboulaye, in his *Moral and Political Studies*, says,
"The new-born Republic just missed dying in its

cradle.'' It became at the very outset clear that there
was a North and a South ; and it was apparent that
the admission of new States must as far as possible be
alternately made in each section, so as to leave a supposed balance of power undisturbed. It was not unfortunate that the question of government, based on
sympathy either with English or with French antece-
dents, developed rapidly two parties,—the Federal and
the Anti-federal, the centralizing and the distributive.
Factionalism counterbalanced, and in some degree neu-
tralized, sectionalism. In 1788, while the people were
waiting for what was called the '' new government,''
Kentucky was once more threatening. Congress had
deferred the recognition of this western territory as an
independent State until the Constitution could be set in
operation. Their delegate went back disgusted and
angry, believing '' they would immediately separate
from the Union.'' Hamilton, when urging his Funding and Assumption Bills, professed to be convinced
that those measures were necessary '' to prevent the
creditor States from seceding from the debtor States.''

Hamilton had laid before the Constitutional Convention eleven propositions which he would make the basis
of the new government. These were so monarchical in
tone that he received no support whatever. As a consequence, he left the Convention, and remained away
from its deliberations. Referring afterward to the
accepted Constitution, he said : '' No man's ideas are
more remote from the plan than my own are known to
be ; but is it possible to deliberate between anarchy
and convulsion on the one side, and the chance of
good from this plan on the other side ? '' Jay had
declined to sign the Declaration of Independence, and
had always remained a very conservative democrat.

He stood for a class of men who would not have been
surprised at a failure of the attempt at popular government. Such men as these formed the core of the
Federal party. Among them were not a few who were
more than doubtful of the Republic. Hamilton declared
that the most appropriate name for the new nation
would be "a Federal Monarchy." In March, 1798,
King, of New York, and Cabot, of Massachusetts, did
not hesitate to say that if the appropriation bills failed
to pass they would "throw up the game." Dexter was
an avowed believer that the presidency must be changed
to the life tenure. Hamilton seems never to have hesitated to repeat his lack of faith in the Constitution and
the government. In 1791, he said : " I own it is my
opinion, though I do not publish it from Dan to Beer-
sheba, that the present government is not that which
will answer the end of society, and that it will be found
expedient to go into the British form." But he did
publish these opinions farther than from Dan to Beer-
sheba. Franklin, Madison, and Jefferson, among the
great leaders of the hour, seemed to be alone in their
entire faith in popular government. Washington, with
superb poise of character, stood between sections and
between factions. "Washington," says Jefferson in a
letter to Martin Van Buren, " was sincerely a friend to
the republican principle of our Constitution. His faith,
perhaps, might not have been as confident as mine, but
he repeatedly declared to me that he was determined
the Constitution should have a fair chance for success,
and that he would lose the last drop of his blood in its
support, against any attempt which might be made to
change its republican form."

The peculiar character of the French Revolution,
with its atrocious development of brutality, added

greatly to the difficulty of the situation in America. The parallel was too close. Both nations labored in the name of human freedom. Both professed to be demonstrations of popular government. The enemies of republicanism had only to cry out Jacobin to create prejudice. The doubters feared that the American people also would drift into excesses. The Whiskey Rebellion added counsel to timidity. A few wished themselves safely back under the English Constitution. An illustration of the influence of the French Revolution was happily touched off by an eminent Virginian with these parallel epitaphs :

Beneath this stupendous pile,	Beneath this horrid pile,
Reared by the hands of mil-	Raised by the hands
lions,	Of
A	Democratic rage
Monster lies,	And
Of millions once the dread	Maniac fury,
and pest,	A consecrated victim
Of Tyranny	Lies ;
The parent and the child ;	Of scepter's royalty
Of Slavery	The
The offspring and the nurse ;	Holy progeny ;
Of mad Ambition	Hyperion's form
The pamper'd minion ;	Was not more fair,
Of Liberty	More luminous,
The scourge and victim.	Or
A	Majestic.
Hypogriff	Of thrones
In form was not more terrible ;	The brightest ornament
A harpy	And
Not more vile, rapacious,	Bulwark ;
Or	Of
Insatiate.	Order, honour, loyalty,
The shores of fertile Nile,	And
Nor all the fields	Chivalry
Of harvest-waving Sicily	The sacred archetype ;

Nor all the cates
That earth or ocean yield,
Could ever satiate
His luxury or appetite.
Ægeon's heads
And
Hundred hands,
To his
Were few.
Not
Typhon,
Nor
Enceladus
So many acres overspread;
Nor could
Huge Ætna's monstrous base
Afford
A grave sufficient.

Reader!
Would'st thou enquire
What Hercules
This monster slew?
'T was LIBERTY!
His grave
Is
France!
On whose broad base
This
Glorious fabric
Securely rests!
Would'st thou know by whom
'Twas rear'd?
The French nation.
Would'st thou
Ask
The monster's name?
It was
NOBILITY! ! !

Of
Anarchy,
The curb and scourge.

A demoniac band
Of spirits, issuing from the
vasty deep,
Sprung, sudden, up,
To immolate
The
Heaven-born Sire;
And,
At Confusion's shrine,
With
The anointed head,
The martyr son
Laid low.

Reader!
Would'st thou enquire
In what devoted clime
This
Pandemonium
Rose?
In France!
Would'st thou give a name
To this
Infernal legion?
The national assembly!
Would'st thou
Canonize
The Saint-like victim?
It was
NOBILITY! ! !

It is impossible to overestimate the uncertain footing of the new government. From 1788 to 1800 was a purely experimental era. Washington said with pathos, "Every step I take is on new ground." There was no precedent in history for such a Federal Union of Independent States. What could be suggested by the Achæan and Hanseatic Leagues was rather misleading than helpful. Had the events of the first ten years of national life been foreseen, it is certain that Virginia and New York, with North Carolina, and probably Pennsylvania, would not have ratified the Constitution. It is equally certain that could New England have looked far enough ahead to have seen the events of the ten years following 1800, it would never have entered the Union.

Nullification had been the rule under the Confederacy. Each State decided for itself whether to respond to Congress. Under the new Union this habit was not easily forsaken. The Empire State was still imperial in spirit and tone, and by itself entered on a system of internal improvement that would have staggered the general government. Massachusetts seemed to Governor Hancock fully as sovereign as ever. Washington found it necessary to make a tour of the States, partly that the people might be accustomed to conceive of a national government as a visible fact. In Boston, the governor insisted that he must himself receive the first visit of state. Washington did not yield the point. When Hancock found the President was about to leave the city, he realized that the game was up ; and having had himself swathed in flannels, he was carried on the shoulders of four men to pay the respects of Massachusetts to the general government. It was a ludicrous but not an unimportant matter. The excuse of illness

2

was sufficient, so long as the State did not establish its
sovereignty over the Union.   Hancock illustrated
State jealousy still more markedly by sending a special
message to the Massachusetts legislature, protesting
against the right of Congress to require him as governor
to certify to the electoral lists made out after a Presidential election.

The effort of eleven States to break loose from the
Union in 1860–61 was not an episode dependent on a
novel reading of Constitutional rights, nor was it solely
a consequence of the desire to perpetuate a social system based on slavery.   It is a very partial and a very
partisan reading of American history that fails to see
that from the acceptance of the Constitution in 1790,
there has been a tendency to assert the right of States
to nullify national enactments or even to sever their
relations to the Union.   This has been a shifting sentiment ; asserted now at the South and then at the
North.   Overt acts have been six in number.   The
first of these occurred in 1798, and in Virginia and
Kentucky took the shape of Nullification Resolutions.
The second was the effort of New England, in 1803,
to create a Northern Confederacy, consisting of five
New England States with New York and New Jersey.
The third was the desperate effort of Vice-President
Burr to create a cleavage in the Southwest, including
the Mississippi Valley, and, hopefully, Ohio.   The
fourth in the disagreeable list was the practical withdrawal of the New England States from co-operation in
the war of 1812–14 ; ending in a Convention of those
States to formulate sectional autonomy.   The fifth act
was in the form of nullification, and was confined to
South Carolina.   The sixth and final act was that of
1861, when eleven States withdrew their representa-

tives from Washington, and created a distinct Confederacy. We may concede that in all these cases the result involved a wholesome discussion of federal problems.

It is my purpose to write with impartiality, as an American and not as a Northerner, the history of these six attempts at Nullification and Secession in the United States. My desire is to make more clear not only historic facts but the dangers that beset a federated republic. It must become with the American people a fixed principle so to conduct public affairs that no section shall desire to be alien to the commonwealth ; or if the desire should arise, can show reasonable cause for overt action. In all cases government is valuable only as it promotes liberty ; as liberty is useless without government. A prosperous people cannot be so except as a whole. Sectional jealousies and suspicions, and all ungenerous sentiments cannot be too thoroughly discouraged.

In a preliminary way it will be useful to note that American history has developed in three characteristic, but somewhat overlapping eras ;—that of faction extending from the forming of the Constitution to about 1816 ; that of sectionalism covering the period from 1820 to 1860 ; and, thirdly, that of a struggle of labor with capital which set in immediately after the war. During the first, the citizens of the States were divided by their sentiments of friendship for either England or France ; New England being in general anxious for closer English alliance, while the Middle and Southern States were inclined to favor measures friendly to France. The second war with England practically closed the struggle of these factions, and began a more independent self-centred national life. The sharp hate of the second or sectional period, bitter as it was, and ending in civil war, did not equal the virulence of the popular hate which characterized the wrangling of Federals and Anti-federals from 1789 to the Hartford Convention of 1814. The present era of centralization, protected trusts, and general State socialism is more dangerous than were the battles of factions and of sections. With a growth of centralized power in combination with capital, a new Bourbonism arises, that the people will be slow to attack, because too long dazzled with the growth of great enterprises, vast cities, and the show of prosperity.

APPENDIX TO CHAPTER I

NOTE WRITTEN BY THOMAS JEFFERSON

The legislature of Virginia happened to be in session when news was received of the passage, by the British Parliament, of the Boston Port Bill. The House of Burgesses thereupon passed a resolution recommending to their fellow citizens that that day should be set apart for fasting and prayer to the Supreme Being, imploring Him to avert the calamities then threatening us, and to give us one heart and one mind to oppose every invasion of our liberties. It was agreed also by the meeting that the Burgesses who should be elected under the writs then issuing, should be requested to meet in convention on a certain day in August ; and to appoint delegates to a congress, should that measure be approved by the other colonies.

INSTRUCTIONS FOR THE DEPUTIES APPOINTED BY VIRGINIA, IN 1774, TO MEET IN GENERAL CONGRESS

The unhappy disputes between Great Britain and her American colonies, which began about the third year of the reign of his present Majesty ; and since, continually increasing, have proceeded to lengths so dangerous and alarming, as to excite just apprehensions in the minds of his Majesty's faithful subjects of this colony, that they are in danger of being deprived of their natural, antient, constitutional, and chartered rights, have compelled them to take the same into their most serious consideration ; and, being deprived of their usual and accustomed mode of making known their grievances, have appointed us their representatives, to consider what is proper to be done in this dangerous crisis of American affairs. It being our opinion, that the united wisdom of North America, should be collected in a General Congress of all the colonies, we have appointed the Honorable Peyton Randolph, Richard Henry Lee, George Washington, Patrick Henry, Richard Bland, Benjamin Harrison, and Edmund Pendleton, Esquires, deputies, to represent this colony in the said Congress, to be held at Philadelphia, on the first Monday in September next.

And that they may be the better informed of our sentiments, touching the conduct we wish them to observe on this important occasion, we desire that they will express, in the first place, our faith and true allegiance to his Majesty, King George the Third, our lawful and rightful sovereign ; and that we are determined, with our lives and fortunes, to support him in the legal exercise of all his just rights and prerogatives. And, however misrepresented, we sincerely approve of a constitutional connection with Great Britain, and wish, most ardently, a return of that intercourse of affection and commercial connection, that formerly united both countries, which can only be effected by a removal of those causes of discontent, which have of late unhappily divided us.

It cannot admit of a doubt, but that British subjects in America, are entitled to the same rights and privileges, as their fellow subjects possess in Britain ; and therefore, that the power assumed by the British Parliament, to bind America by their statutes, in all cases whatsoever, is unconstitutional, and the source of these unhappy differences.

The end of government would be defeated by the British Parliament exercising a power over the lives, the property, and the liberty of American subjects ; who are not, and, from their local circumstances, cannot be, there represented. Of this nature, we consider the several acts of Parliament, for raising a revenue in America, for extending the jurisdiction of the courts of Admiralty, for seizing American subjects, and transporting them to Britain, to be tried for crimes committed in America, and the several late oppressive acts respecting the town of Boston, and Province of the Massachusetts Bay.

The original constitution of the American colonies, possessing their assemblies with the sole right of directing their internal polity, it is absolutely destructive of the end of their institution, that their legislatures should be suspended, or prevented, by hasty dissolutions, from exercising their legislative powers.

Wanting the protection of Britain, we have long acquiesced in their acts of navigation, restrictive of our commerce, which we consider as an ample recompense for such protection ; but as those acts derive their efficacy from that foundation alone, we have reason to expect they will be restrained, so as to produce the reasonable purposes of Britain, and not be injurious to us.

To obtain redress of these grievances, without which the people of America can neither be safe, free, nor happy, they are willing to undergo the great inconvenience that will be derived to them, from stopping all imports whatsoever, from Great Britain, after the first day of November next, and also to cease exporting any commodity whatsoever, to the same place, after the tenth day of August, 1775. The earnest desire we

have to make as quick and full payment as possible of
our debts to Great Britain, and to avoid the heavy
injury that would arise to this country, from an earlier
adoption of the non-exportation plan, after the people
have already applied so much of their labor to the
present crop, by which means, they have been prevented from pursuing other methods of clothing and
supporting their families, have rendered it necessary to
restrain you in this article of non-exportation ; but it
is our desire, that you cordially co-operate with our
sister colonies in General Congress, in such other just
and proper methods as they, or the majority, shall deem
necessary, for the accomplishment of these valuable
ends.

BILL OF RIGHTS

Passed by the Continental Congress, October 14, 1774.

The Congress came into the following Resolutions :
Whereas, since the close of the last war, the British
parliament, claiming a power of right to bind the people of America, by statute in all cases whatsoever, hath
in some acts expressly imposed taxes on them, and in
others under various pretences, but in fact for the purpose of raising a revenue, hath imposed rates and
duties payable in these colonies, established a board of
commissioners with unconstitutional powers, and extended the jurisdiction of courts of admiralty, not only
for collecting the said duties, but for the trial of causes
merely arising within the body of a county.

And whereas, in consequence of other statutes, judges,
who before held only estates at will in their offices,
have been made dependent on the Crown alone for
their salaries; and standing armies kept in time of

peace. And whereas it has lately been resolved in Parliament, that by force of statute, made in the thirty-fifth year of the reign of King Henry the eighth, colonists may be transported to England and tried there upon accusations for treasons and misprisions, or concealments of treasons committed in the colonies; and by late statute, such trials have been directed in cases therein mentioned.

And whereas, in the last session of Parliament, three statutes were made one; entitled, "An act to discontinue in such manner, and for such time as are therein mentioned, the landing and discharging, lading or shipping of goods, wares and merchandise, at the town, and within the harbor of Boston, in the province of Massachusetts-Bay, in North America." Another entitled, "An act for the better regulating the government of the Province of the Massachusetts-Bay, in New-England." And another, entitled, "An act for the impartial administration of justice, in the cases of persons questioned for any act done by them in the execution of the law, or for the suppression of riots and tumults, in the Province of the Massachusetts-Bay, in New-England." And another statute was then made, " for making more effectual provision for the government of the Province of Quebec, &c." All which statutes are impolitic, unjust, and cruel, as well as unconstitutional, and most dangerous and destructive of American rights.

And whereas, assemblies have been frequently dissolved, contrary to the rights of the people, when they attempted to deliberate on grievances; and their dutiful, humble, loyal, and reasonable petitions to the crown for redress, have been repeatedly treated with contempt by his Majesty's ministers of state.

The good people of the several colonies of New-
Hampshire, Massachusetts - Bay, Rhode - Island and
Providence plantations, Connecticut, New-York, New-
Jersey, Pennsylvania, New Castle Kent and Sussex
sembled, in a full and free representation of these col-
on Delaware, Maryland, Virginia, North Carolina, and
South-Carolina, justly alarmed at these arbitrary proceedings of parliament and administration, have severally elected, constituted, and appointed deputies to
meet and sit in general congress, in the city of Philadelphia, in order to obtain such establishment, as that
their religion, laws, and liberties may not be subverted.
Whereupon the deputies so appointed, being now assembled, in a full and free representation of these colonies, taking into their most serious consideration the
best means of attaining the ends aforesaid, do in the
first place, as Englishmen, their ancestors, in like cases
have usually done, for asserting and vindicating their
rights and liberties, Declare,

That the inhabitants of the English colonies in
North America, by the immutable laws of nature, the
principles of the English constitution, and the several
charters or compacts, have the following Rights :

Resolved, N. C. D.   1. That they are entitled to life,
liberty, and property ; and they have never ceded to
any sovereign power whatever a right to dispose of
either without their consent.

Resolved, N. C. D.   2. That our ancestors, who first
settled these colonies, were at the time of their emigration from the mother country, entitled to all the rights,
liberties, and immunities of free and natural born subjects, within the realm of England.

Resolved, N. C. D.   3. That by such emigration they
by no means forfeited, surrendered, or lost any of those

rights, but that they were, and their descendants now are, entitled to the exercise and enjoyment of all such of them, as their local and other circumstances enable them to exercise and enjoy.

Resolved, 4. That the foundation of English liberty, and of all free government, is a right in the people to participate in their legislative council : and as the English colonists are not represented, and from their local and other circumstances cannot properly be represented in the British parliament, they are entitled to a free and exclusive power of legislation in their several provincial legislatures, where their right of representation can alone be preserved, in all cases of taxation and internal polity, subject only to the negative of their sovereign, in such manner as has been heretofore used and accustomed : But from the necessity of the case, and a regard to the mutual interests of both countries, we cheerfully consent to the operation of such acts of the British parliament, as are, bona fide, restrained to the regulations of our external commerce, for the purpose of securing the commercial advantages of the whole empire to the mother country ; and the commercial benefits of its respective members, excluding every idea of taxation, internal or external, for raising a revenue on the subjects in America without their consent.

Resolved, N. C. D. 5. That the respective colonies are entitled to the common law of England, and more especially to the great and inestimable privilege of being tried by their peers of the vicinage, according to the course of the law.

Resolved, 6. That they are entitled to the benefit of such of the English statutes as existed at the time of their colonization ; and which they have, by experience, respectively found to be applicable to their several local and other circumstances.

Resolved, N. C. D. 7. That these his Majesty's colonies are likewise entitled to all the immunities and privileges granted and confirmed to them by royal charters, or secured by their several codes of provincial ·laws.

Resolved, N. C. D. 8. That they have a right peaceably to assemble, consider of their grievances, and petition the King ; and that all prosecutions, prohibitory proclamations, and commitments for the same, are illegal.

Resolved, N. C. D. 9. That the keeping a standing army in these colonies, in times of peace, without the consent of the legislature of that colony in which such army is kept, is against law.

Resolved, N. C. D. 10. It is indispensably necessary to good government, and rendered essential by the English constitution, that the constituent branches of the legislature be independent of each other ; that, therefore, the exercise of legislative power in several colonies, by a council appointed, during pleasure, by the crown, is unconstitutional, dangerous, and destructive to the freedom of American legislation.

All and each of which, the aforesaid deputies, in behalf of themselves, and their constituents, do claim, demand, and insist on, as their indubitable rights and liberties ; which cannot be legally taken from them, altered or abridged, by any power whatever, without their own consent, by their representatives in their several provincial legislatures.

In the course of our inquiry, we find many infringements and violations of the foregoing rights ; which, ·from an ardent desire that harmony and mutual intercourse of affection and interest may be restored, we pass over for the present, and proceed to state such acts and measures as have been adopted since the last war, which demonstrate a system formed to enslave America.

Resolved, N. C. D. That the following acts of parliament are infringements and violations of the rights of the colonists ; and that the repeal of them is essentially necessary, in order to restore harmony between Great Britain and the American colonies, viz. :

The several acts of 4 Geo. III. ch. 15 and ch. 34 :—5 Geo. III. ch. 25 :—6 Geo. III. ch. 52 :—7 Geo. III. ch. 41 and ch. 46 :—8 Geo. III. ch. 22 ; which impose duties for the purpose of raising a revenue in America ; extend the powers of the admiralty courts beyond their ancient limits ; deprive the American subject of trial by jury ; authorize the judge's certificate to indemnify the prosecutor from damages, that he might otherwise be liable to, requiring oppressive security from a claimant of ships and goods seized, before he shall be allowed to defend his property, and are subversive of American rights.

Also 12 Geo. III. ch. 24, entitled, "An act for the better securing his Majesty's dock-yards, magazines, ships, ammunition, and stores." Which declares a new offence in America, and deprives the American subject of a constitutional trial by jury of the vicinage, by authorizing the trial of any person, charged with the committing any offence described in the said act out of the realm, to be indicted and tried for the same in any shire or county within the realm.

Also the three acts passed in the last session of Parliament, for stopping the port and blocking up the harbor of Boston, for altering the charter and govern-

ment of Massachusetts-Bay, and that which is entitled,
" An act for the better administration of justice," &c.

Also the act passed in the same session for establishing the Roman Catholic religion in the Province of
Quebec, abolishing the equitable system of English
laws, and erecting a tyranny there, to the great danger, from so total a dissimilarity of religion, law, and
government, to the neighboring British colonies, by
the assistance of whose blood and treasure the said
country was conquered from France.

Also the act passed in the same session, for the better
providing suitable quarters for officers and soldiers in
his Majesty's service in North-America.

Also that the keeping a standing army in several of
these colonies in times of peace, without the consent of
the legislature of that colony in which such army is
kept, is against law.

ARTICLES OF ASSOCIATION, OCTOBER 20, 1774

We his Majesty's most loyal subjects, the Delegates
of the several Colonies of New-Hampshire, Massachusetts-Bay, Rhode-Island, Connecticut, New-York,
New-Jersey, Pennsylvania, the Three Lower Counties
of Newcastle Kent and Sussex on Delaware, Maryland, Virginia, North-Carolina, and South-Carolina,
deputed to represent them in a continental Congress,
held in the city of Philadelphia, on the fifth day of
September, 1774, avowing our allegiance to his Majesty,
our affection and regard for our fellow subjects in
Great-Britain and elsewhere, affected with the deepest
anxiety, and most alarming apprehensions at those
grievances and distresses, with which his Majesty's
American subjects are oppressed, and having taken

under our most serious deliberation, the state of the whole continent, find, that the present unhappy situation of our affairs, is occasioned by a ruinous system of colony-administration adopted by the British Ministry about the year 1763, evidently calculated for enslaving these Colonies, and, with them, the British Empire. In prosecution of which system, various Acts of Parliament have been passed for raising a Revenue in America, for depriving the American subjects, in many instances, of the constitutional trial by jury, exposing their lives to danger, by directing a new and illegal trial beyond the seas, for crimes alleged to have been committed in America ; and in prosecution of the same system, several late, cruel, and oppressive Acts have been passed respecting the town of Boston and the Massachusetts-Bay, and also an Act for extending the province of Quebec, so as to border on the western frontiers of these Colonies, establishing an arbitrary government therein, and discouraging the settlement of British subjects in that wide-extended country ; thus by the influence of civil principles and ancient prejudices to dispose the inhabitants to act with hostility against the free Protestant Colonies, whenever a wicked Ministry shall choose so to direct them.

To obtain redress of these grievances, which threaten destruction to the lives, liberty, and property of his Majesty's subjects in North-America, we are of opinion, that a non-importation, non-consumption, and non-exportation agreement, faithfully adhered to, will prove the most speedy, effectual, and peaceable measure :— And therefore we do, for ourselves, and the inhabitants of the several Colonies whom we represent, firmly agree and associate under the sacred ties of virtue, honor, and love of our country, as follows :

First. That from and after the first day of December next, we will not import into British America, from Great-Britain or Ireland, any goods, wares or merchandize whatsoever, or from any other place any such goods, wares or merchandize, as shall have been exported from Great Britain or Ireland ; nor will we, after that day, import any East-India tea from any part of the world ; nor any molasses, syrups, paneles, coffee, or pimento, from the British plantations, or from Dominica ; nor wines from Madeira, or the Western-Islands ; nor foreign indigo.

Second. That we will neither import, nor purchase any slave imported after the first day of December next, after which time we will wholly discontinue the slave-trade, and will neither be concerned in it ourselves, nor will we hire our vessels, nor sell our commodities or manufactures to those who are concerned in it.

Third. As a non-consumption agreement, strictly adhered to, will be an effectual security for the observation of the non-importation, we, as above, solemnly agree and associate, that, from this day, we will not purchase or use any tea imported on account of the East-India Company, or any on which a duty hath been or shall be paid ; and from and after the first day of March next, we will not purchase or use any East-India tea whatever, nor will we, nor shall any person for or under us, purchase or use any of those goods, wares or merchandize, we have agreed not to import, which we shall know, or have cause to suspect, were imported after the first day of December, except such as come under the rules and directions of the tenth article hereafter mentioned.

Fourth. The earnest desire we have, not to injure

our fellow subjects in Great-Britain, Ireland or the West-Indies, induces us to suspend a non-exportation until the tenth day of September, 1775; at which time, if the said Acts and parts of Acts of the British Parliament, hereinafter mentioned, are not repealed, we will not, directly or indirectly, export any merchandize or commodity whatsoever, to Great-Britain, Ireland, or the West-Indies, except rice to Europe.

Fifth. Such as are merchants, and use the British and Irish trade, will give orders, as soon as possible, to their factors, agents, and correspondents, in Great-Britain and Ireland, not to ship any goods to them, on any pretence whatsoever, as they cannot be received in America ; and if any merchant, residing in Great-Britain or Ireland, shall directly or indirectly ship any goods, wares or merchandize, for America, in order to break the said non-importation agreement, or in any manner contravene the same, on such unworthy conduct being well attested, it ought to be made public ; and on the same being so done, we will not from thenceforth have any commercial connexion with such merchant.

Sixth. That such as are owners of vessels will give positive orders to their captains, or masters, not to receive on board their vessels any goods prohibited by the said non-importation agreement, on pain of immediate dismission from their service.

Seventh. We will use our utmost endeavors to improve the breed of sheep, and increase their number to the greatest extent, and to that end we will kill them as sparingly as may be, especially those of the most profitable kind ; nor will we export any to the West-Indies, or elsewhere ; and those of us who are or may become overstocked with, or can conveniently spare

3

any sheep, will dispose of them to our neighbors, especially of the poorer sort, on moderate terms.

Eighth. That we will in our several stations encourage frugality, economy, and industry ; and promote agriculture, arts, and the manufactures of this country, especially that of wool ; and will discountenance and discourage, every species of extravagance and dissipation, especially all horse-racing, and all kinds of gaming, cock-fighting, exhibitions of shows, plays, and other expensive diversions and entertainments. And on the death of any relation or friend, none of us, or any of our families, will go into any further mourning dress, than a black crape or ribbon on the arm or hat for Gentlemen, and a black ribbon or necklace for Ladies, and we will discontinue the giving of gloves and scarfs at funerals.

Ninth. That such as are venders of goods or merchandize, will not take advantage of the scarcity of goods that may be occasioned by this association, but will sell the same at the rates we have been accustomed to do, for twelve months last past.—And if any vender of goods or merchandize, shall sell any such goods on higher terms, or shall in any manner, or by any device whatsoever, violate or depart from this Agreement, no person ought, nor will any of us deal with any such person, or his, or her factor or agent, at any time thereafter, for any commodity whatever.

Tenth. In case any merchant, trader, or other persons shall import any goods or merchandize after the first day of December, and before the first day of February next, the same ought forthwith at the election of the owner, to be either re-shipped or delivered up to the committee of the county, or town wherein they shall be imported, to be stored at the risk of the importer,

until the non-importation agreement shall cease, or be
sold under the direction of the committee aforesaid ;
and in the last-mentioned case, the owner or owners of
such goods, shall be reimbursed (out of the sales) the
first cost and charges, the profit, if any, to be applied
towards relieving and employing such poor inhabitants
of the town of Boston, as are immediate sufferers by the
Boston Port-Bill ; and a particular account of all goods
so returned, stored, or sold, to be inserted in the public
papers, and if any goods or merchandizes shall be imported after the said first day of February, the same
ought forthwith to be sent back again, without breaking any of the packages thereof.

Eleventh.   That a committee be chosen in every
county, city, and town, by those who are qualified to
vote for Representatives in the Legislature, whose business it shall be attentively to observe the conduct of all
persons touching this association; and when it shall be
made to appear to the satisfaction of a majority of any
such committee, that any person within the limits of
their appointment has violated this association, that
such majority do forthwith cause the truth of the case
to be published in the Gazette, to the end that all such
foes to the rights of British America may be publicly
known, and universally contemned as the enemies of
American liberty ; and thenceforth we respectively will
break off all dealings with him or her.

Twelfth.   That the committee of correspondence in
the respective colonies do frequently inspect the entries
of their custom-houses, and inform each other from time
to time of the true state thereof, and of every other
material circumstance that may occur relative to this
association.

Thirteenth.   That all manufactures of this country

be sold at reasonable prices, so that no undue advantage be taken of a future scarcity of goods.

Fourteenth. And we do further agree and resolve, that we will have no trade, commerce, dealings, or intercourse, whatsoever, with any colony or province, in North-America, which shall not accede to, or which shall hereafter violate this association, but will hold them as unworthy of the rights of freemen, and as inimical to the liberties of their country.

And we do solemnly bind ourselves and constituents, under the ties aforesaid, to adhere to this association until such parts of the several Acts of parliament passed since the close of the last war, as impose or continue duties on tea, wine, molasses, syrups, paneles, coffee, sugar, pimenta, indigo, foreign paper, glass, and painters colors, imported into America, and extend the powers of the Admiralty courts beyond their ancient limits, deprive the American subject of trial by jury, authorize the judge's certificate to indemnify the prosecutor from damages, that he might otherwise be liable to from a trial by his peers, require oppressive security from a claimant of ships or goods seized, before he shall be allowed to defend his property, are repealed—And until that part of the Act of the 12 Geo. 3. ch. 24 entitled, "An Act for the better securing his Majesty's dock-yards, magazines, ships, ammunition, and stores," by which, any persons charged with committing any of the offences therein described, in America, may be tried in any shire or county within the realm, is repealed—And until the four Acts passed in the last session of Parliament, viz: That for stopping the port and blocking up the harbor of Boston— That for altering the charter and government of the Massachusetts-Bay—And that which is entitled, "An Act for the better administration of justice, &c."— And that "For extending the limits of Quebec, &c." are repealed. And we recommend it to the provincial conventions, and to the committees in the respective Colonies, to establish such farther regulations as they may think proper, for carrying into execution this Association.

The foregoing Association being determined upon by the Congress, was ordered to be subscribed by the several Members thereof ; and thereupon we have hereunto set our respective names accordingly.

In Congress, Philadelphia, October 20, 1774.

Signed, PEYTON RANDOLPH, President,

and by the Members.

JEFFERSON'S ORIGINAL DRAFT OF THE DECLARATION OF INDEPENDENCE. REPORTED JULY 4TH AND SIGNED JULY 9th, 1776 ; AS AMENDED

When, in the course of human events, it becomes necessary for one people to dissolve the political bands which have connected them with another, and to assume among the powers of the earth, the separate and equal station to which the laws of nature and of nature's God entitle them, a decent respect to the opinions of mankind requires that they should declare the causes which impel them to the separation.

We hold these truths to be self-evident : that all men are created equal ; that they are endowed by their creator with inherent and inalienable rights ; that among these are life, liberty, and the pursuit of happiness ; that to secure these rights, governments are instituted among men, deriving their just powers from the consent of the governed ; that whenever any form of government becomes destructive of these ends, it is the right of the people to alter or to aboilsh it, and to institute new government, laying its foundation on such principles, and organizing its powers in such form, as to them shall seem most likely to effect their safety and happiness.   Prudence, indeed, will dictate that governments long established should not be changed for light and transient causes ; and accordingly all experience hath shewn that mankind are more disposed to suffer while evils are sufferable, than to right themselves by abolishing the forms to which they are accustomed.   But when a long train of abuses and usurpations, begun at a distinguished period, and pursuing invariably the same object, evinces a design to reduce them under absolute despotism, it is their right, it is their duty to throw off such government, and to provide new guards for their future security.   Such has been, the patient sufferance of these colonies ; and such is now the necessity which constrains them to expunge their former systems of government.   The history of the present king of Great Britain is a history of unremitting injuries and usurpations ; among which appears no solitary fact to contradict the uniform tenor of the rest, but all have in direct object the establishment of an absolute tyranny over these states.   To prove this, let facts be submitted to a candid world, for the truth of which we pledge a faith yet unsullied by falsehood.

He has refused his assent to laws the most wholesome and necessary for the public good.

He has forbidden his governors to pass laws of immediate and pressing importance, unless suspended in their operation till his assent should be obtained ; and,

when so suspended, he has utterly neglected to attend
to them.

He has refused to pass other laws for the accommodation of large districts of people, unless those people
would relinquish the right of representation in the
legislature, a right inestimable to them, and formidable
to tyrants only.

He has called together legislative bodies at places
unusual, and distant from the depository of their public records, for the sole purpose of fatiguing them into
compliance with his measures.

He has dissolved representative houses repeatedly
and continually, for opposing with manly firmness his
invasions on the rights of the people.

He has refused for a long time after such dissolutions
to cause others to be elected, whereby the legislative
powers, incapable of annihilation, have returned to the
people at large for their exercise ; the state remaining,
in the meantime, exposed to all the dangers of invasion
from without and convulsions within.

He has endeavored to prevent the population of these
states ; for that purpose obstructing the laws for naturalization of foreigners, refusing to pass others to
encourage their migrations hither, and raising the conditions of new appropriations of lands.

He has suffered the administration of justice totally
to cease in some of these states ; refusing his assent to
laws for establishing judiciary powers.

He has made our judges dependant on his will alone
for the tenure of their offices, and the amount and payment of their salaries.

He has erected a multitude of new offices, by a self-
assumed power, and sent hither swarms of new officers
to harass our people and eat out their substance.

He has kept among us in times of peace standing armies and ships of war without the consent of our legislatures.

He has affected to render the military independent of, and superior to the civil power.

He has combined with others to subject us to a jurisdiction foreign to our constitutions and unacknowledged by our laws, giving his assent to their acts of pretended legislation for quartering large bodies of armed troops among us ; for protecting them by a mock trial from punishment for any murders which they should commit on the inhabitants of these states ; for cutting off our trade with all parts of the world ; for imposing taxes on us without our consent ; for depriving us of the benefits of trial by jury ; for transporting us beyond seas to be tried for pretended offences ; for abolishing the free system of English laws in a neighboring province, establishing therein an arbitrary government, and enlarging its boundaries, so as to render it at once an example and fit instrument for introducing the same absolute rule into these states ; for taking away our charters, abolishing our most valuable laws, and altering fundamentally the forms of our governments ; for suspending our own legislatures, and declaring them invested with power to legislate for us in all cases whatsoever.

He has abdicated government here, withdrawing his governors, and declaring us out of his allegiance and protection.

He has plundered our seas, ravaged our coasts, burnt our towns, and destroyed the lives of our people.

He is at this time transporting large armies of foreign mercenaries to complete the works of death, desolation and tyranny, already begun with circumstances of cruelty and perfidy unworthy the head of a civilized nation.

He has constrained our fellow citizens taken captive on the high seas, to bear arms against their country, to become the executioners of their friends and brethren, or to fall themselves by their hands.

He has endeavored to bring on the inhabitants of our frontiers the merciless Indian savages, whose known rule of warfare is an undistinguished destruction of all ages, sexes, and conditions of existence.

He has incited treasonable insurrections of our fellow citizens with the allurements of forfeiture and confiscation of our property.

He has waged cruel war against human nature itself, violating its most sacred rights of life and liberty in the persons of a distant people who never offended him, captivating and carrying them into slavery in another hemisphere, or to incur miserable death in their transportation thither.   This piratical warfare, the opprobrium of INFIDEL powers, is the warfare of the CHRISTIAN king of Great Britain.   Determined to keep open a market where MEN should be bought and sold, he has prostituted his negative for suppressing every legislative attempt to prohibit or to restrain this execrable commerce.   And that this assemblage of horrors might want no fact of distinguished die, he is now exciting those very people to rise in arms among us, and to purchase the liberty of which he has deprived them, by murdering the people on whom he also obtruded them: thus paying off former crimes committed against the LIBERTIES of one people, with crimes which he urges them to commit against the LIVES of another.

In every stage of these oppressions we have petitioned

for redress in the most humble terms : our repeated petitions have been answered only by repeated injuries.

A prince whose character is thus marked by every
act which may define a tyrant, is unfit to be the ruler
of a people who mean to be free. Future ages will
scarcely believe that the hardiness of one man adven-
tured, within the short compass of twelve years only,
to lay a foundation so broad and so undisguised for
tyranny over a people fostered and fixed in principles
of freedom.

Nor have we been wanting in attentions to our British brethren. We have warned them from time to time
of attempts by their legislature to extend a jurisdiction
over these our states. We have reminded them of the
circumstances of our emigration and settlement here,
no one of which could warrant so strange a pretension :
that these were effected at the expense of our own
blood and treasure, unassisted by the wealth or strength
of Great Britain : that in constituting indeed our several forms of government, we had adopted one common
king, thereby laying a foundation for perpetual league
and amity with them : but that submission to their
parliament was no part of our constitution, nor ever in
idea, if history may be credited : and, we appealed to
their native justice and magnanimity, as well as to the
ties of our common kindred to disavow these usurpations, which were likely to interrupt our connection
and correspondence. They too have been deaf to the
voice of justice and of consanguinity, and when occasions have been given them, by the regular course of
their laws, of removing from their councils the disturbers of our harmony, they have by their free election, reestablished them in power. At this very time
too, they are permitting their chief magistrate to send
over not only soldiers of our common blood, but Scotch

and foreign mercenaries to invade and destroy us. These facts have given the last stab to agonizing affection, and manly spirit bids us to renounce forever these unfeeling brethren. We must endeavor to forget our former love for them, and hold them as we hold the rest of mankind, enemies in war, in peace friends. We might have been a free and a great people together ; but a communication of grandeur and of freedom, it seems, is below their dignity. Be it so, since they will have it. The road to happiness and to glory is open to us too. We will tread it apart from them, and acquiesce in the necessity which denounces our eternal separation !

We therefore the representatives of the United States of America in General Congress assembled, do in the name, and by authority of the good people of these states, reject and renounce all allegiance and subjection to the kings of Great Britain, and all others who may hereafter claim by, through or under them ; we utterly dissolve all political connection which may heretofore have subsisted between us and the people or parliament of Great Britain : and finally we do assert and declare these colonies to be free and independent states; and that as free and independent states, they have full power to levy war, conclude peace, contract alliances, establish commerce, and to do all other acts and things which independent states may of right do.

And for the support of this declaration, we mutually pledge to each other our lives, our fortunes, and our sacred honor.

DIGEST OF THE ARTICLES OF CONFEDERATION

Article I.—The style of this Confederacy shall be, " The United States of America."

Art. II.—Each State retains its sovereignty, freedom, and independence, and every power, jurisdiction, and right, which is not by this Confederation expressly delegated to the United States in Congress assembled.

Art. III.—The said States hereby severally enter into a firm league of friendship with each other, for their common defence, the security of their liberties, and their mutual and general welfare, binding themselves to assist each other against all force offered to, or attacks made upon them, or any of them, on account of religion, sovereignty, trade, or any other pretense whatever.

Art. IV.—The better to secure and perpetuate mutual friendship and intercourse among the people of the different States in this Union, the free inhabitants of each of these States, paupers, vagabonds, and fugitives from justice excepted, shall be entitled to all privileges and immunities of free citizens in the several States ; and the people of each State shall have free ingress and egress to and from any other State, and shall enjoy therein all the privileges of trade and commerce subject to the same duties, impositions, and restrictions as the inhabitants thereof respectively ; provided that such restrictions shall not extend so far as to prevent the removal of property imported into any State to any other State of which the owner is an inhabitant ; provided also, that no imposition, duties, or restriction shall be laid by any State on the property of the United States or either of them.   If any person guilty of, or charged with, treason, felony, or other high misdemeanor in any State shall flee from justice and be found in any of the United States, he shall, upon the demand of the governor or executive power of the State from which he fled, be delivered up and removed to the

State having jurisdiction of his offence. Full faith
and credit shall be given in each of these States to the
records, acts, and judicial proceedings of the courts
and magistrates of every other State.

Art. VI.— . . . No two or more States shall
enter into any treaty, confederation, or alliance whatever between them, without the consent of the United
States, in Congress assembled, specifying accurately
the purposes for which the same is to be entered into,
and how long it shall continue.

Art. IX. . . . The United States, in Congress
assembled shall be the last resort on appeal in all disputes and differences now subsisting, or that hereafter may arise between two or more States concerning
boundary, jurisdiction, or any other cause whatever;
which authority shall always be exercised in the manner following : Whenever the legislative or executive
authority, or lawful agent of any State in controversy
with another, shall present a petition to Congress, stating the matter in question, and praying for a hearing,
notice thereof shall be given by order of Congress to
the legislative or executive authority of the other State
in controversy, and a day assigned for the appearance
of the parties by their lawful agents ; who shall then
be directed to appoint by joint consent, commissioners
or judges to constitute a court for hearing and deter-
mining the matter in question ; but if they cannot
agree, Congress shall name three persons out of each
of the United States, and from the list of such persons
each party shall alternately strike out one, the peti-
tioners beginning, until the number shall be reduced

to thirteen ; and from that number not less than seven
or more than nine names, as Congress shall direct,
shall, in the presence of Congress, be drawn out by lot ;
and the persons whose names shall be so drawn, or any
five of them shall be commissioners or judges, to hear
and finally determine the controversy.

. . . The United States in Congress assembled,
shall have authority to appoint a committee, to sit in
the recess of Congress to be denominated " A Committee of the States," and to consist of one delegate from
each State ; and to appoint such other committees and
civil officers as may be necessary for managing the
general affairs of the United States under their direction ; to appoint one of their number to preside, provided that no person be allowed to serve in the office
of president more than one year in any term of three
years.

Art. X.—The Committee of the States, or any nine
of them, shall be authorized to execute in the recess of
Congress, such of the powers of Congress as the United
States, in Congress assembled, by the consent of nine
States, shall, from time to time, think expedient to in-
vest them with ; provided that no power be delegated to
the said Committee, for the exercise of which, by the
Articles of Confederation, the voice of nine States in
the Congress of the United States assembled is requisite.

Art. XI.—Canada, acceding to this Confederation,
and joining in the measures of the United States, shall
be admitted into, and entitled to all the advantages of
this Union ; but no other colony shall be admitted into
the same, unless such admission be agreed to by nine
States.

.   .   .   .

Article XIII.— . . . And the Articles of this
Confederation shall be inviolably observed by every
State, and the Union shall be perpetual.

Done at Philadelphia, July 9, 1778.

HAMILTON'S PLAN OF GOVERNMENT; LAID BEFORE
THE CONSTITUTIONAL CONVENTION OF 1788

1. The supreme legislative power of the United
States of America to be vested in two distinct bodies of
men ; the one to be called the assembly, the other the
senate ; who, together, shall form the legislature of the
United States ; with power to pass all laws whatsoever,
subject to the negative hereafter mentioned.

2. The assembly to consist of persons elected by the
people, to serve for three years.

3. The senate to consist of persons elected to serve
during good behavior ; their election to be made by
electors chosen for that purpose by the people. In
order to this, the states to be divided into election districts. On the death, removal, or resignation of any
senator, his place to be filled out of the district from
which he came.

4. The supreme executive authority of the United
States to be vested in a governor, to be elected to serve
during good behavior. His election to be made by
electors, chosen by electors, chosen by the people in
the election districts aforesaid. His authorities and
functions to be as follows :—

To have a negative upon all laws about to be passed,
and the execution of all laws passed ; to have the entire
direction of war, when authorized, or begun ; to have,

with the advice and approbation of the senate, the power of making all treaties ; to have the sole appointment of the heads or chief officers of the departments of finance, war, and foreign affairs ; to have the nomination of all other officers (ambassadors to foreign nations included) subject to the approbation or rejection of the senate ; to have the power of pardoning all offences, except treason, which he shall not pardon, without the approbation of the senate.

5. On the death, resignation, or removal of the governor, his authorities to be exercised by the president of the senate, until a successor be appointed.

6. The senate to have the sole power of declaring war; the power of advising and approving all treaties ; the power of approving or rejecting all appointments of officers, except the heads or chiefs of the departments of finance, war, and foreign affairs.

7. The supreme judicial authority of the United States to be vested in —— judges, to hold their offices during good behavior, with adequate and permanent salaries. This court to have original jurisdiction in all causes of capture ; and an appellative jurisdiction in all causes, in which the revenues of the general government, or the citizens of foreign nations, are concerned.

8. The legislature of the United States to have power to institute courts in each state, for the determination of all matters of general concern.

9. The governors, senators, and all officers of the United States to be liable to impeachment for mal and corrupt conduct ; and, upon conviction, to be removed from office, and disqualified for holding any place of trust, or profit. All impeachments to be tried by a court to consist of the chief, or senior judge of the superior court of law in each state ; provided, that such

judge hold his place during good behavior, and have a permanent salary.

10. All laws of the particular states, contrary to the constitution or laws of the United States, to be utterly void. And the better to prevent such laws being passed, the governor or president of each state shall be appointed by the general government, and shall have a negative upon the laws about to be passed in the state of which he is governor, or president.

11. No.state to have any forces, land or naval ; and the militia of all the states to be under the sole and exclusive direction of the United States ; the officers of which to be appointed and commissioned by them.

4

CHAPTER II

THE first problem before the new nation was who should decide whether the general government kept faith with the States. Could it decide on its own act, or should the States themselves pass judgment as to whether more power was exercised than they had granted? It is not surprising that the contracting parties claimed that they, having made the government, were judges of what the government might do. Hamilton was quick to see that the only hope for a strong central government was to assert its right to pass on its own actions. The essays constituting the Federalist were along this line ; and were among the ablest political papers ever produced. Some of these were written by Hamilton, others by Madison, and by those who afterwards became Republicans.

There was a deal of thinking going on, and fortunately for the young nation the civil service had not become subordinated to party fealty ; so that the people had the continuous aid of the ablest men they could produce. Imagine the supreme disaster if Thomas Jefferson could not have been in almost continuous public service from the first years of his mature life.

The result was that Franklin, Jefferson, Madison, Hamilton, Washington, Adams, while moulding the Republic, could themselves grow into larger views and learn by trial and experience. In fact they felt their way through dangers not now comprehensible. The natural drift of minds was in the two directions of equality, and of aristocracy. Those who felt strongest the Saxon heredity believed English institutions were, apart from monarchy, very closely ideal. Those who had also some comprehension of individualism, de-veloped more largely in France, were ready to dispense with aristocracies and trust the people. Hamilton and Jefferson soon became leaders of the two schools of politics. They were masters well pitted. Hamilton's eleven propositions in the Constitutional Convention were indicative of his tendencies of thought. Jefferson can be best comprehended by a study of the Declara-tion of Independence.

The confidence that Jefferson had in the people as contrasted with any form of aristocracy, follows a well-known law that the ablest commoners have been of noble blood. This has been true in England, France, Germany, and Russia as well as in the United States. Jefferson was of the Randolph stock, as proud a family as Virginia held ; but of such pedigree he merely re-marked; " Any one might ascribe to it the faith and merit he chose." As for himself he cared nothing for " cyphers of aristocracy." He was a member of the House of Burgesses at twenty-six. At thirty-one he wrote the *Summary View of the Rights of British America* ; in which he told King George that " kings are the servants, not proprietors of the people ; and that the whole art of government is the act of being honest." At thirty-three he wrote the Declaration of Independence, and later was author of the Ordinance excluding slavery from our Northwestern Territories.   Not afraid of experiments ; a mind of extraordinary synthetic powers ; an idealist who fully believed in the possible realization of the Golden Rule ; cautious in action but prompt in a crisis, Jefferson in politics was what Franklin was in physics, a student of the past but not a slave to precedent.   Hamilton was rash in action but conservative in theory.   Capable of adjusting himself to conditions that he disapproved, he never rid himself of aristocratic sentiments and a lack of confidence in the people.   By nature an intriguer, both in his social relations and in public affairs, he not seldom weakened his cause by intricate plots to overthrow his opponents.

Jefferson says of his experience in Washington's cabinet, " Hamilton and I were pitted against each other every day in the cabinet, like two fighting-cocks."   Washington stood not only between the two to moderate and soften antagonism, but to prevent a collision of their followers.   It was destined that American history, down to the present time at least, should be a conflict of Hamiltonian and Jeffersonian ideas and methods.   Washington found more help in immediate stress of events from Hamilton.   Jefferson looked more into the future, laying deep and broad the basis of the educational and political and social structure.

Hamilton's Funding law was joined with an Assumption law, that, transferring all State war debts to the nation, increased the opposition to both measures. Virginia memorialized Congress to repeal them.   Hamilton, never able to comprehend opposition, declared that such a spirit " must be stamped out."   There was no doubt truth in Jefferson's charge that, while Hamilton accepted the Constitution, he did so with the belief that

it would be temporary ; and he intended to see that it
was temporary.   As late as 1802 he wrote, " I am still
trying to prop the frail and worthless fabric."   On another occasion he averred: " I myself am affectionately
attached to the republican theory, and desire to demonstrate its practical success.   But as to State governments, if they can be circumscribed consistently with
preserving the Union, it is well.   I seriously apprehend
that the United States will not be able to maintain itself
against their influence.   Hence I am disposed for a
liberal construction of the powers of the general government."   He urged that there be an amendment of the
Constitution, granting Congress power to cut up the
States at its option into smaller ones.   Such a plan
would of course have given the party in power perpetuity of rule.   Jefferson grew so suspicious of these
theories and plans that he wrote, taken altogether,
their object in his judgment was " to draw all the
powers of government into the hands of the general
legislature; to establish means for corrupting a sufficient
corps in that legislature, to subvert step by step the
principles of the Constitution."

   " Where," said Jefferson, " do they lead us but to
the overthrow of the Republic and the establishment of
a monarchy."   To the taunt " Jacobin " the Jeffer-
sonians   hurled   back   the   epithet   " Monarchists."
Monroe in 1817, referring to this era, wrote to Andrew
Jackson: " That some of the leaders of the Federal party
entertained principles unfriendly to our system of government I have been thoroughly convinced ; that they
meant to work a change in it by taking advantage of
favorable circumstances I am equally satisfied."   But
he thought these purposes were never known by any
large number of the rank and file of the Federal party.

The leaders dare not trust the people ; the people too fully trusted the leaders.

That the Hamilton party should first come to the front was inevitable for several reasons. The sympathies of Washington were in favor of considerable dignity and appanage. But the weakness of the old Confederacy had left a widespread fear of a second attempt at government without power. The war left the people astounded with their own independence. They had to create a nation, whether they desired to do it or not. It was natural to look for precedents. England after all was mother of America. The people were accustomed to her policy, socially and politically. The Federal party consisted of those who leaned heavily on English tradition.

The church and the state went together. Both were Anglican in sympathy. Intensely hating the free thought that was on the increase in the States, and attributing this to French influence, the clergy were nearly unanimous, not only for the party that stood for the established, but as unanimous in antagonism to those who began to style themselves Republicans. Successful in electing not only Washington, but his successor ; with no political warnings to guide them or foretell disaster ; the Federal party pushed matters to extremes. Their intent was evident, not only to conduct public affairs for the time being, but to control the policy of the country in the future. To accomplish this required of them to crush their antagonists ; and this they proceeded to do. A Federal Congress legislated against Republicans as " Jacobins " ; Federal courts packed juries to try them ; and judges delivered political harangues to insure their conviction.

The war between England and America was by no

means ended by the treaty of peace. Each charged the
other with breaches of the compact. England insisted
that the tories had been outraged ; and hindrances put
in the way of collecting debts due British subjects.
The Americans answered,—You carried off thousands
of our negroes, you have not surrendered the western
posts as bound to do by treaty ; and you are using those
posts to incite the Indians to hostility. Adams, who
was our minister at London in 1785, wrote that the
treaty of Paris had only gained this for us that we could
now fight " without halters around our necks."
" There is," he wrote to Jay, " no chance for anything
but retaliations, reciprocal prohibitions, and imposts."
England hated us all the worse for having been forced
to yield our independence. In 1786 Adams wrote in
disgust, that both England and France were planning
to get all they could out of the United States. He
added, " I cannot get rid of the persuasion that the fair
plant of liberty in America must be watered in blood."

It was this state of affairs that made Washington
anxious to negotiate a commercial treaty with England,
and bring the strain to a relaxation, if not to a satisfactory conclusion. Yet it must be conceded that the
first political disaster that befel the young nation was
the Jay treaty of 1795. Although we had compelled
Great Britain to acknowledge our independence, our
common and national rights were continually the subject of insult and robbery. Jay it was supposed had
been sent out to assert our rights and defend our honor.
But when the treaty was received it was kept locked
up till the Senate could be called together to act on it.
That body then saw fit to meet with closed doors. Protests came in from every quarter. But the Federal
infatuation with England was not to be hindered by

protest. The treaty was ratified by a vote of 20 to 10,
although its contents even then were not made known
to the public.   But they soon leaked out in part ; and
then one of the Senators, Mr. Mason, gave his copy to
the press.   The country was at once in a rage of resentment.   Indignation meetings were called in Boston,
Philadelphia, Baltimore, and New York.   Hamilton,
haranguing a crowd in defence of the treaty, was
stoned.   Address after address was rushed to President
Washington to stay his signature.   It was in fact a
strange document for a free even though a feeble people
to consent to.   The twelfth article forbade American
vessels to carry from America or from English ports to
Europe any coffee, cocoa, sugar, molasses or cotton.   It
did nothing to stop impressment of our seamen and
intrusion upon our vessels.   It gave Great Britain the
right to use all our rivers and ports ; but it shut us out
of Canadian waters.   England had illegally retained
certain posts that she had pledged herself to surrender
at the close of the war.   This treaty gave her a year
and a half more of possession.   Not the least objection
to the treaty was that France, which had stood by us in
our hour of trial was aggrieved, and her interests outraged as badly as our own, for the sake of peace with
Great Britain.   The people of Savannah burned Jay in
effigy.   In Charleston the treaty was burned by the
hangman.   If the people had before been friendly to
France they were now wildly enthusiastic for their
former allies.   The heart of the masses beat true to
gratitude.

The treaty was interpreted as a desire to establish an
aristocracy after the British pattern.   The Federal
party was denounced as a half-way house to monarchy.
That Washington should have come in for some of this

almost universal indignation is not to be wondered at. Ramsay says that Washington signed the treaty with great hesitation, and only because, if it were rejected, there seemed to him no alternative left but war. Jefferson wrote to Edmund Rutledge, " I join with you in thinking the treaty execrable." As " Camillus," Hamilton entered the list in defence of the offensive document. France, with whom we held a treaty of alliance, abrogated the same without delay. It was in fact the beginning of a long series of disasters. From France we could thereafter expect only indignation and resentment, while England was more domineering than ever. The inevitable end of such a treaty must be war, and so it proved. It confined us in a position unendurable for any length of time ; and gave Great Britain a treaty-right to domineer and insult us. The only apology for the Jay treaty ever put in print was the general one that "It was calculated to remove a variety of causes of uneasiness, of complaint, of recrimination." John Adams wrote, " It certainly cannot be ranked a triumph of American diplomacy, but it is better than war."

Making all possible allowances for the exigencies of the time, the Jay treaty was a party treaty. Mr. Jay himself wrote to Judge Cushing : " I was well-convinced that it would not receive anti-federal approbation." It was signed by the commissioners November 19, 1794. On May 28, 1795, Jay was in New York. The treaty was not published until July 2, the day after Jay was inaugurated governor. Although ratified in secret by the Senate, June 24th, it was not signed by Washington till August 15, 1795. His just and reasonable dread was another war ; yet he delayed signature for two months ; making nearly a year from

its signature by the commissioners. And not till March
3, 1796, was it proclaimed as supreme law. The storm of
exasperated citizenship that broke out when the treaty
was published did not end till the War of 1812 ; when
both countries welcomed the chance to get rid of what
remained of its articles. Overbearing England had
granted more than she had intended ; victorious
America had found herself bound by a treaty to hold
inferior rank among nations.

John Adams succeeded Washington in 1796. Allowing that he had the most severe task ever assigned to
any President—not even excepting Lincoln,—and well
seeing that the young nation had no precedent, and a
vast future—everything to learn, and no history to
warn or guide it,—yet it is inexplicable, looking back
from our present vantage ground of judgment, why the
Federalist party of his administration should have
shown so little political wisdom, and carried so high a
hand.

Adams was no more than installed before Hamilton
issued an address,—" message," Adams called it, to
Congress, advising, with the passage of Alien and Sedition laws, the raising of an army of fifty thousand men.
The President was astounded, and refused to endorse
either measure. Adams wrote, " He graciously conde-
scended to delineate a system of administration, to project negotiations, and nominate ambassadors."

The clash came at once. Hamilton had been consulted by Washington in every affair, however minute.
Adams was not so easily managed. The quarrel culminated in a virulent but shrewd pamphlet by Hamilton. This was based on material dishonorably obtained
by him from Adams's disloyal cabinet ; on the whole a
most extraordinary combination of meanness and

treachery. Adams had generously and confidingly continued in office Washington's rump cabinet. These men, Pickering, Wolcott, and McHenry, in particular, belonged to Hamilton ; and continuously not only thwarted their chief, but betrayed him. They had been accepted by Washington, rather than selected ; for after the resignations of Jefferson, Hamilton, and Knox, the highest offices had gone begging. The annoyance to Washington had been so very great it was thought that he himself contemplated resigning. It was a lesson in American politics. Adams proved to be thoroughly independent ; nor is there any adequate evidence that he was fractious or guided by jealousy. He was self-willed; while his policy was cautious, wise, and well-informed.

Slighting the recommendations of the President, Congress proceeded to pass upon those of Hamilton. An Alien law was passed, and a Sedition law, followed by a vital amendment to the naturalization laws. But instead of an army of fifty thousand as Hamilton desired, ten thousand of which were to be cavalry, they ordered that thirteen thousand men should be armed and equipped as a provisional force.

Adams objected to the army proposed by Hamilton as " the wildest extravagance of a knight errant." " He observed," he said, " that an unqualified recommendation of war with France was expected of him." He as unqualifiedly refused to recommend war. "Pernicious," wrote Adams, " it was to Hamilton's views of ambition and domination. It extinguished his hopes of being at the head of an army of fifty thousand men ; without which he used to say he had no idea of having a head on his shoulders for four years longer." Adams further insisted that Hamilton should leave the seat of

government and go back to Newark, to drill his troops
—" if capable of it."

The first draft of the Alien and Sedition laws, and
the Amendment of the Naturalization laws, excluding
foreigners for sixteen years before naturalization, would
have killed off most of the voters from American citizenship, and were of such a tyrannous nature that Hamilton himself suggested modification.   The Sedition Act
made it felonious and punishable with a fine of $5000,
.and five years' imprisonment, for persons to combine in
order to impede the operation of any law of the United
States, or to intimidate persons from taking federal
office, or to commit or advise a riot, or insurrection, or
unlawful assembly.   It also declared that the writing or
publishing of any scandalous, malicious, or false statement against the President, or either house of Congress,
should be punishable by a fine of $2000 and imprison-
ment for two years.   The bill as first presented in the
Senate, made it treason " to give aid or comfort to the
people of France," who were pronounced to be " enemies of the United States " ; and it punished by fine
and imprisonment " any person who should justify the
course pursued by the French, or who should defame
the existing government of the United States by
declarations tending to create a belief that any officers
of the government were influenced by motives hostile
to the Constitution or the liberties of the people."
These clauses of the Bill were stricken out ; but as
finally passed it remained an act of dangerous usurpation.   It left it impossible freely to discuss an Act of
Congress.   It was a death-blow to freedom of speech
and of the press.   Livingston declared the Act would
have " disgraced Gothic barbarism."

But infinitely worse was the Alien Act, passed by the

same Congress, at the same session. Under plea of civil commotion, resulting from the residence of foreigners in the States, it was enacted that any alien might be summarily banished, without assigned cause, and at the sole discretion of the President. There need have been no suit, nor even public charges. Any suspect might be ordered to depart, within specified time, from the limits of our Union ; and if he failed to obey, he might on order of the President be imprisoned for three years. Imprisoned aliens might be deported if the executive so chose ; and masters of vessels must always make report, on arrival, of all aliens on board. These acts were seconded by a proposition of the Massachusetts legislature to amend the Constitution, prohibiting from service in Congress all foreign-born citizens, except those resident in the United States at the time of the Declaration of Independence. Otis, the brilliant representative of Boston Federalism, offered a resolution that no foreign-born person should hereafter hold office in the United States. It was probably provoked by the fact that Gallatin, a Swiss, at this time was rapidly rising to eminent leadership of the Republicans.

Of the Alien and Sedition laws Adams, as late as 1813, boasted to Jefferson that the former was never executed by him in a single instance. But in a letter to Pickering in 1799 he certainly did recommend the deportation of an obnoxious editor. Pickering in return suggested getting rid of Dr. Priestley; and implied in correspondence that Adams gave him authority to expel a certain General Collot.

But if the Alien Acts were not rigidly enforced by President Adams, the Sedition Act was enforced. The outrages on personal liberty were as serious as any complained of under Royal colonial governors. The prosecutions passed from the oppressive to the grotesque. President Adams, travelling through Newark, was complimented by the firing of cannon. A man named Baldwin said he wished the wadding might hit the President's backsides. For this he was arrested, and had to pay one hundred dollars. Frothingham, an editor, was imprisoned and fined for saying that Hamilton '' attempted to buy the Aurora in order to suppress it.'' Thomas Cooper was kept in prison six months and fined four hundred dollars for saying that President Adams in 1797 was '' hardly in the infancy of political mistakes.'' Ten printers and editors were among those prosecuted for sedition. Among these were B. F. Bache, Thomas Cooper, and Abijah Adams. A new moral as well as criminal code was created. In the case against Callender, Judge Chase harangued against the accused, and ordered the marshal to put none but Federalists on the jury. On Callender's lawyers he threatened to inflict corporal punishment. Judge Peck, of Otsego, N. Y., circulated a petition for the repeal of the offensive acts. For this action he was indicted, and taken to New York for trial. Matthew Lyon, Congressman from Vermont, canvassing for re-election, charged the President with '' unbounded thirst for ridiculous pomp, foolish adulation, and a selfish avarice.'' He was thrust into jail, and held for four months ; besides paying a fine of one thousand dollars. Sent by his constituents back to Congress, an attempt was made to prevent him taking his seat. Such was the determination of the party in power to perpetuate its rule and to suppress criticism. In 1840, Congress refunded to Lyon's heirs the amount of his fine, with interest.

Popular protest against the acts of the Congress of '98 grew rapidly in force. Indignation passed into alarm. Virginia in her General Assembly adopted resolutions calling on the States to nullify the enforcement of the Sedition laws. The resolutions declared that Virginia promised support to the government of the United States in all measures warranted by the Constitution ; but declared the acts of the general government " no farther valid than they are authorized by the grants enumerated in that compact ; and that in case of a deliberate, palpable, and dangerous exercise of other powers, not granted by said compact, the States who are parties thereto have the right, and are in duty bound, to interpose for arresting the progress of the evil, and for maintaining within their respective limits the authorities, rights, and liberties appertaining to them." This general assembly thereupon declared by law of the State of Virginia, that the Alien and Sedition laws were unconstitutional ; and invited other States to co-operate in resisting them ; " in order to maintain unimpaired the authorities, rights, and liberties reserved to the States respectively, or to the people." Madison himself drafted these resolutions ; and they expressed not only his views, but those of Jefferson, and Mason, and Patrick Henry, and of nearly every other Virginian of eminence.

The action of Virginia was seconded by that of her territorial child, Kentucky. This was done not only at the suggestion of Jefferson himself, but in the main the Resolutions were the work of his hand. They declared that where powers were assumed by the national government which had not been granted by the States, " nullification is the rightful remedy "; and that every State has an original natural right " to nullify of its own authority all assumptions of power by others, within its limits." The States it was hoped would

recur to their natural right, in all cases not made federal, " to take measures that they shall not be exercised within their respective territories."

The provocation was extreme, as every citizen of the
United States will now allow. And we shall not be far
apart in our judgment of the value of a Union that
would permit such arbitrary exercise of power. Citizenship would be intolerable. Only a class in power would
be safe from confiscation, and immolation or banish-
ment. Exercised to the possible limit by any administration, it would give it a perpetual lease of power ; and
work as thoroughly as a French *coup d'état*. Kentucky thought that the national government could not
justly be the exclusive or final judge of the extent of
the powers delegated to itself ; since " that would make
its discretion and not the Constitution the measure of
its powers."

The exposition of these nullifying resolutions, as
given by Jefferson himself, shows that it was not by
any means desired to weaken the Federal Union ; but
to express a general right of the States to prevent unjust and unconstitutional assumptions of Congressional
power. If the Federal party would begin with the
Alien and Sedition Acts, where would it end ? Jefferson argued that if two or three States should secede,
and form a separate compact, " the same difficulties
might occur in the smaller union ; and finally each
unit fall apart into its colonial condition." He might
have added that the end might not be there ; but
that each State was liable to disruption by the secession of counties. In a letter concerning the Kentucky
Resolutions, he says he desired to leave the matter in
such a train as that they should " not be committed
absolutely to push the matter to extremities ; and yet

free to push as far as the events render prudent." It
is certain however, that Jefferson had become fully
alarmed at the encroachments of power, and had made
up his mind that liberty must be sustained at all cost.
Virginia built a great armory at Richmond, and prepared for whatever contingency might occur.   Adams
justly termed the position, not only of Virginia and
Kentucky but of " the whole South and West, as
menacing."

Other States saw differently.   The Assembly of New
York resolved that " as the right of deciding on the
constitutionality of laws passed by Congress pertains to
the judiciary department of government, this House
disclaims the power assumed in and by the resolutions
of the legislatures of Virginia and Kentucky."   Other
States responded in a similar tone.   This was due in
large degree to the influence of Hamilton and his Federalist.   The issue was joined which was to have remote consequences.   Had the contracting parties—the
States—lost by the terms of the contract all right to
judge of the honesty with which the contract was ful-
filled ?

The position of Virginia and Kentucky on the question of fealty to the Union being challenged by New
York and New England, they responded near the close
of 1799, by a new set of resolutions.   Kentucky declared unqualified loyalty to the Constitution ; but
declined positively to assent to the doctrine that the
general government was sole judge of its own acts ; " a
doctrine that must end sooner or later in absolute despotism."   Both States insisted that palpable violations
of the Constitution ought to be nullified by the States.
To the Constitution Kentucky would bow unqualifiedly ; to any other authority only with qualification.

5

The merits of this controversy divided into : (1) **Who** were the contracting parties to the formation of the United States ; (2) Who should decide whether laws enacted by the national legislature were constitutional, or extra constitutional.

(1) Notwithstanding the ingenious and useful argument of Webster, that the people created the nation, it still remains true that, in doing so, they acted in their organic capacity as States.   The Constitution opens with the Preamble that, " We the *people* of the United States in order to form a more perfect Union, establish justice, insure domestic tranquility, provide for the common defense, promote the general welfare and secure the blessings of liberty to ourselves and our posterity, do ordain and establish this Constitution for the United States of America."   But it closes with Article VII, which is, " The ratification of the Conventions of nine *States* shall be sufficient for the establishment of this Constitution, between the *States* so ratifying the same."   And Washington set his name after the words, " Done in convention by the unanimous consent of the States present."   In the Constitutional Convention the voting was by States ; each State having one vote.   Article X of the Amendments to the Constitution proposed by the Congress of 1789 (adopted 1791) reads, " The powers not delegated to the United States by the Constitution, nor prohibited by it to the States, are reserved to the States respectively or to the people."   It is open to question whether the words States and people in this clause are used as synonyms, or as distinguishing some reserved powers as belonging to the people not taken organically.

Nor must it be forgotten that the Constitution, if it had been passed upon by the people directly, would not have been adopted. Chief Justice Marshall, himself a strong Federalist, allows that "so small in many instances was the majority for the Constitution, as to afford strong ground for the opinion that had the influence of character been removed, the intrinsic merits of the instrument would not have secured its adoption. Indeed it is scarcely to be doubted that in some of the States a majority of the people were in opposition." It is therefore a case of *suppressio veri* to say that the people were the contracting parties in constituting the nation, while it is allowed that if they had voted directly on the Constitution, it would not have been adopted. It was the people acting as States that constructed the Federal Union. No State exists except as an organ of the people ; but as such it is as truly an organ as the head or hand or foot is an organ of the individual.

In the ratifying acts of two States the right of the States to withdraw from the Union was distinctly asserted. Virginia said, "That the powers granted under the Constitution, being derived from the people of the United States, may be resumed by them, whensoever the same shall be perverted to their injury or oppression ; and that every power not granted thereby, remains with them and at their will." This seems to be not only an implied right of the States to nullify, but to withdraw from the Union.

And yet Mr. Marshall in a Supreme Court decision of 1816 says: "The Convention which framed the Constitution was indeed elected by the State legislatures. But the instrument when it came from their hands was a mere proposal, without obligation, or pretensions to it. : . . By the Convention and by Congress and by the State legislatures the instrument was submitted to the people. They acted upon it in the only manner in which they can act, by assembling in convention. It is true they assembled in States, but where else should they have assembled ? . . . The assent of the States in their sovereign capacity is implied in calling a convention; and thus submitting that instrument to the people. But the people were at perfect liberty to accept or reject it, and their act was final. It requires not their affirmance, and could not be negatived by the State governments.'' This zigzag use of facts amounts only to this, that the people as States created, and afterwards as States adopted the Constitution. The confusion of trying to distinguish between State governments and the people is absurd ; for the town can be conceived only as organized people ; and the State as organized towns or their equivalents. There is no State government apart from the people. But Federalism, filled with British political instinct, found it impossible not to conceive the government as a distinct entity. This mischievous misconception permeates our history everywhere to this day, to our serious damage. It is not at all necessary to deny Mr. Marshall's final assertion that " the government of the Union is emphatically and truly a government of the people," in order to affirm that, beyond contradiction, our government is also a compact of the people in their organic capacity as States. The evil of nullification must be based elsewhere in order to meet the facts of history and the conditions of the case. We have only to amend the allowance that " It is true they assembled in States," by saying they assembled *as* States.

If however the Supreme Court, or Mr. Webster, or Andrew Jackson, finds it necessary to insist on a distinction between " the people " and " the State " or " the government," then we must refer to another chapter of history. The people's Convention of New Jersey, ratifying the Constitution, said, " Whereas, the legislature of this State did pass an act to *authorize* the people of the State to meet in convention, deliberate upon, agree to and ratify the Constitution of the United States, we the delegates of the State of New Jersey do hereby ratify, &c." Connecticut said in convention, " We the delegates of the people of said State, pursuant to an act of the legislature, do ratify, &c." Virginia met " in pursuance of a recommendation of the General Assembly." It not only ratified the Constitution as it was, but recommended an amendment " That each *State* in the Union shall respectively retain every power, jurisdiction and right which is not by this Constitution delegated to the Congress of the United States." But Virginia as we have seen reserved the right to nullify or withdraw from the Union ; and in doing so asserted it as the right of the·people. The ' abstract distinction between " people " and " State governments " was hatched by the fertile brain of Federalism.

But this doctrine never took shape as a general principle until formulated by Mr. Webster in his debate with Senator Hayne in 1832. He then laid it down as a fundamental principle that the Constitution is not a compact between the people of the several States in their sovereign capacities, but a government proper ; founded on the adoption of the people, and creating direct relations between itself and individuals. But Mr. Webster himself had said in his reply to Mr. Foote in 1830: " I am resolved not to submit in silence to accusations which impute to us a disposition to evade the Constitutional compact." John Quincy Adams impatiently commented that " the Union is both of the people and of

the States ; and that all constitutional government is a
compact.'' Calhoun quoted Webster against himself
with great force. The latter thereupon modified his
position to '' the Constitution *rests* on compact, and is
no longer a compact; for if it be a compact States may
secede.''

The history of the ratifying conventions contains
further evidence that the government was considered
to be a compact of sovereign States. In the Massachusetts convention Fisher Ames argued that '' The
Senators represent the sovereignty of the States.''
John Sumner added '' The General Government depends on the State governments for its very existence; ''
and Judge Parsons declared '' The government is the
mere result of a compact.'' '' The resolutions of ratification define the Constitution as a solemn compact.'' The
same understanding is definitely incorporated in the
discussions and ratifying resolutions of other States.
No fact is more certain than that the Constitution
would never have been accepted on any other basis.

This view of the States, as corporate members of a
Union by compact, was common throughout the North
during the contest over the Fugitive Slave Law.
Senator Wade of Ohio said in Congress in 1855 : '' Who
is to be the judge, in the last resort, of the violation of
the Constitution of the United States, by the enactment
of a law ? Who is the final arbiter ? The general
government ? or the States in their sovereignty ? Why,
sir! to yield that point is to yield all the rights of the
States to protect their citizens ; and to consolidate this
government into a miserable despotism.'' Again : '' I
said there were States in this Union whose highest
tribunals had adjudged that bill to be unconstitutional ;
and that I was one of those who believed it unconsti-

tutional ; and that under the old resolutions of 1798 a State must not only be the judge of that, but of the remedy in such case."

There is great danger in drawing the distinction between States and people.   For if the people can act upon Federal affairs in any other capacity than by States, in State convention, or State assembly, by delegates, then we open the way to the assertion on the part of any section or any class to avow a law higher than Congress. Indeed Jackson did assert this doctrine in one of his veto messages.   "Each public officer," he said, "takes an oath to support the Constitution as he understands it ; and not as it is understood by others."   It was a very similar sentiment expressed by Channing and Wm. H. Seward in the assertion of a "higher law." This meant clearly individual right to hold conscientious convictions above the decisions of the Supreme Court, and acts of Congress as well.   If the general government is not a compact of States, there has been no gain in the structure of government;—federalism is a dream.   The end will be centralization ; weakening of the States ; with ultimate dissolution of State sentiment and Statehood.

(2) It does not follow from what has been shown in the preceding section, that the Supreme Court is not to decide upon the constitutionality of the laws passed by Congress.   But is its decision final and conclusive ?   Up to the time of the Sedition and Alien Acts the place of the Supreme Court had not clearly been apprehended. It was the last one of the three co-ordinate branches of government to grow into shape.   There were up to 1800 less than fifty cases on the docket.   Judges resigned to take positions now held to be vastly less honorable.   It had been created to declare the law in

actual cases of contest. It did not pass on any uncontested act of Congress. But when called upon to thus act, its province was conceded to cover not only a definition of the Constitution, but the laws which were enacted under the Constitution. The right of States to primarily decide on acts of Congress would have led to continual confusion. Hamilton did the country no greater service that in pointing out that " the interpretation of the laws is the proper and peculiar province of the courts. It must therefore belong to them to ascertain the meaning of the Constitution, as well as the meaning of any act proceeding from the legislative body."

But however cautiously the Supreme Court moved in finding its legitimate place in the government, one of its earlier decisions brought about a serious clash of authority. In the case of Chisholm against the State of Georgia, the Court decided that under the Constitution a State could be sued by a citizen of another State. This was so clearly correct, that it was necessary to pass the Eleventh Amendment to the Constitution, which declared in a definite form that " The judicial power shall not be construed to extend to any suit in law or equity, commenced or prosecuted against one of the United States by citizens of another State, or by citizens or subjects of any foreign state." But meanwhile the Court had decided otherwise. Massachusetts and Georgia among others were sued. Massachusetts in legislature convened, resolved not to respond to the suit, and to wholly ignore it. But Georgia, more fiery in the temper of its people, passed an Act subjecting to death " any marshal of the United States, or other person, who should presume to serve any process against that State at the suit of an individual."

Clearly there was an appeal from the Supreme Court in this case to the States. Such an appeal was taken ; and the decision of the Court was negatived by a Constitutional Amendment. The Eleventh Amendment of the Constitution was a needed check and limitation of the power of the Supreme Court. It is no answer to say that abstract justice should allow a suitor to collect honest dues from any party. It is true even that States have in some instances escaped from honest indebtedness by shameful acts of repudiation. Neither Congress nor the President nor the Supreme Court is a moral arbiter of such matters.

In Wisconsin in 1855 a conflict arose under the Fugitive Slave Law. It was a case of positive nullification and defiance combined. The decision of Chief-Justice Taney with the majority of the Court was, (1) A free negro of the African race, whose ancestors were brought to this country and sold as slaves, is not a citizen within the meaning of the Constitution. (2) The judgment of the circuit court is therefore erroneous, as it had no jurisdiction of the controversy between the parties. (3) Scott remains a slave. The law making the Territory of Wisconsin free territory is unconstitutional and void. (4) The Missouri Compromise Act of March, 1820 is unconstitutional and void. This was probably a verifiable opinion—but the final court of appeals was the people ; and they negatived it.

In this case the people acted by States. The principle that an officer is bound "to regard and execute the process of the United States, and not to obey the process of the State authorities," was nullified. In Ohio the Supreme Court of the State found that "no law of Congress can compel a State officer to deliver up an alleged criminal to the governor of Kentucky." The

Supreme Court, says Judge Biddle, pathetically confessed, that the performance of the duty in question was left to depend upon the fidelity of the State Executive to the compact with other States.

When the Civil War broke out, the Supreme Court came in conflict with President Lincoln. Chief-Justice Taney decided that only Congress could suspend the writ of habeas corpus. But Lincoln ordered his officers to disobey the writ of the Court ; and they did it successfully. When Mr. Lincoln instituted a blockade, the Supreme Court would have rendered his work nugatory, only for the votes of three judges appointed by Mr. Lincoln himself.

Not only the States and the Executive have thus negatived the Court, but also Congress. In the violent struggle of reconstruction which followed the War of Secession, Congress, while impeaching the President as surpassing his constitutional rights, arbitrarily forbade the filling of a vacancy in the Supreme Court until the number of Associate Justices was reduced to six. The Constitutional number was nine.

In the case of Worcester against the State of Georgia, in 1833, a conflict was precipitated between the State Court and the United States Court. Worcester was arrested under a State law forbidding a white man to reside within the limits of the Cherokee nation without first obtaining a license. The sentence of imprisonment for four years was reversed by the United States Supreme Court. The State Court refused obedience. When President Jackson was urged to execute the decree of the Judicial Department, he replied : " John Marshall has made his decision ; let him execute it."

The Court has been packed by the Executive to ensure the carrying out of measures. President Jackson,

in order to secure an endorsement of his arbitrary deal-
ing with the United States Bank, placed one of his
cabinet on the bench. During the administration of
President Grant the Legal Tender Act was declared
unconstitutional. But the President finding a man
who, as a State judge, had affirmed the constitutionality
of the same law, selected him to fill a vacancy on the
bench of Associates, and by the help of his vote the decision of the Court was reversed.

Conflicts have in other cases terminated more favorably for the U. S. Court. The impeachment of Judge
Chase in 1805 fell through from a confusion in the
Senate as to the basis of impeachment. The trial put
an end however to the partisan conflict between Court
and Congress. The Judges thereafter were less given
to extra-judicial harangues.

In 1819 the Court decided that Congress had constitutional right to charter its own banks in the States,
contrary to the expressed wish and will of the people
of said States. This decision was nullified as far as
possible in several States. The contest in Ohio was
most protracted and bitter. The State adopted the
very same expedient for killing national banks that was
adopted in 1862 by Congress to kill State banks. It
undertook to tax them to death. In the end the State
was beaten. But we find the legislature acting favorably on resolutions to this effect, that " the doctrines
asserted by the legislatures of Kentucky and Virginia in
1798 were sound and true ; that the General Assembly
asserts and will maintain the right of the States to tax
the business and property of any private corporation
chartered by Congress and doing business in the State ;
and that the General Assembly further protests against
the doctrine that the political rights of the separate

States and their powers as sovereign States can be
settled by the Supreme Court of the United States in
cases between individuals, and in which no State is a
party direct."

So we see that the Supreme Court has not been the
final arbiter in questions of Congressional action ; nor
has it been the final authority as to the constitutionality
of such action. The real final authority in our government has been the common sense of the people acting
as States. That the people have rectified not only
Legislative, but Legal blundering, is an historic fact.
They correct or negative Congressional work with
promptitude ; but have learned to be more patient with
the Court.

Jefferson, in a letter dated September 11, 1804, says :
" You seem to think it devolved on the Judges to decide
the validity of the Sedition law. But nothing in the
Constitution has given them a right to decide for the
executive, more than the executive to decide for them.
But the opinion which gives to the Judges the right to
decide what laws are constitutional and what not—not
only for themselves in their own sphere of action, but
for the legislature and executive also, in their spheres,
would make the judiciary a despotic branch." This
plainly brings about a conflict of authority ; the only
possible final court of resort being the people. President Lincoln was supremely a representative of the
nation's accumulated common sense. He said in his
first inaugural : " I do not forget the position assumed
by some that Constitutional questions are to be decided
by the Supreme Court, nor do I deny that such questions must be binding upon the parties to that suit,
while they are also entitled to very high respect and
consideration in all parallel cases by all departments of

the government. But if the policy of the government upon a vital question affecting the whole people, is to be irrevocably fixed by the decisions of the Supreme Court, the moment they are made, the people will have ceased to be their own masters ; having to that extent resigned their government into the hands of that emi-nent tribunal.''

These words of Lincoln are deeply worth the study of those who, since the war, have fallen into the drift to centralize the government, at the expense of the States ; and especially to exalt the Supreme Court.

Admirable illustrations of what President Lincoln avowed were seen not only in the decision concerning legal tender ; but later in the income tax decisions. By this decision an annual income of the government of $30,000,000 was affirmed as constitutional ; by almost the next decision of the same court, within three months, this income was obliterated. In 1870 Secretary of the Treasury John Sherman said that '' the most just and equitable tax levied in the United States is the income tax ; for it is the only discrimination in our tax laws that will reach wealthy men as against the poorer classes who pay nine-tenths of our taxes.'' This honorable distribution of the national burden is now rendered impossible, and all alleviation debarred by the Supreme Court until an appeal can be taken to the people. It is presumed that the Supreme Court is pre-eminently a body of students and investigators ; and therefore more likely to act with deliberate judgment than Congressional legislators. But there was barely thirteen days' notice of the rehearing. More hasty legislation was rarely ever ventured by Congress than these terribly sweeping, but contradictory affirmations of 1895. So '' the Constitution grows '' sometimes smaller and narrower, rather than always larger and fuller, as some are fond of affirming.

We have to go deeper in our nation-building than to establish a court with paramount power, power to sweep away a century at a single sitting. The original conception of the Supreme Court was a body of learned investigators. Its decisions were less important than its examinations of questionable law. And its chief value has ever been to teach the people to wait patiently on its deliberations, instead of rushing into revolutionary action to nullify offensive legislation.

To yield that the Supreme Court has paramount power is to yield farther that it has a right to decide, as it has decided, that Congress has "sovereign power" simply as Congress. Paramount authority lies (1) in the people acting as States ; (2) in the people acting as the United States. The Supreme Court has no paramount power, but only derived and delegated power. It is a body that stands for and represents the people ; and is as responsible to the people as the legislative branch of government.

The peculiar characteristic of the judicial branch of the government, as provided for by the Constitution, was the service of judges for life, unless impeached and removed for misbehavior. This life term has served to separate the judiciary from the other branches of government in spirit and sympathy. Legislators and the President are so closely subject to popular waves of sentiment that they have become watchful of the popular will ; and have adjusted their action to the hour. The Supreme Court, being farther removed from current changes of popular will would be expected to show less fear of the ballot. It may presume too much upon this exemption from frequently being called to the bar

of the people. Hamilton said in the Federalist : " The
judiciary, from the nature of its functions, will always
be least dangerous to the political rights of the Constitution, because it will be least in a capacity to annoy or
injure them. The executive not only dispenses the
honors but holds the sword of the community ; the
legislature not only commands the purse but prescribes
the rules by which the duties and rights of every citizen
are to be regulated ; the judiciary on the contrary has
no influence over the sword or the purse, no direction
of either the strength or the wealth of society, and can
take no active resolution whatever. The judiciary is
beyond question the weakest of the three departments
of power."   And Von Holtz has to add : " There are
besides, other disputed Constitutional questions, which
in their nature can never be brought before the Supreme
Court, or be decided by it.   Moreover, violations of the
Constitution may happen, and the injured cannot,
whether States or individuals, obtain justice through
the Court.   Where the wrongs suffered are political in
origin the remedies must be sought in a political way."
" Yet it does not follow," he thinks, "that the sovereign
States are authorized to interpose."   He does not show
where this power to " authorize " " sovereigns " resides;
or whence it could originate.   It is this difficulty of
bringing the Supreme Court before the bar of the
people that has constituted its danger and weakness as
well as its strength.   For whenever a conflict has arisen,
in almost all cases it has taken some form of revolution.

It was a general suspicion of the judiciary department
that made the acceptance of the Constitution so very
close in 1789.   It was carried in New York only by a
vote of 30 to 27; and in Virginia by 89 to 79. One proposition before the Convention had been to give Congress

power of veto over all State legislation, and a right to
enforce its decisions.   Such a constitutional grant
would have left the States powerless ; and the Resolutions of Virginia and Kentucky would have been direct
revolution.   The present system, says Professor Bur-
gess is " the aristocracy of the robe ; probably," he
adds, " the best for the purpose the world has yet produced."   If it is the best yet devised it is because its
functions undertook to cover the most difficult office of
government.   Our history does not show that this
has been done in a manner entirely satisfactory to the
people.·

Holding to a division of American history into the
periods of a struggle of factions from 1779 to 1816, a
struggle of sections from 1820 to 1860, and a struggle
of classes since 1870, we shall see that the Supreme
Court has in every instance thrown its influence against
popular rights.   In the first era the Supreme Court was
so violently Federal and partisan that two of the judges
were impeached for packing juries, and delivering political harangues while charging juries.   In the second era
the Supreme Court decided that the black man had no
rights as a citizen that a white man was bound to consider.   In the third era the Supreme Court has already
cast its wavering voice, in the social struggle, for the
wealthy as a class.

That the States cannot and will not continue from
time to time to nullify some irrational, or even a just
decision of the Supreme Court, I do not believe.   That
the Executive will find at some future time good cause
to again check or resist the Court is probable.   That
Congress will not show its popular character, and forbid the meddling of the Court is not at all probable.
Contingencies will arise to show that the judiciary is

not in any way above the other departments, or able to enforce authority over either the general government or the States. " The people," says von Holst, " as all the constitutions say, are the sole possessors of political power ; and they alone therefore can give the State its fundamental law. . . . Hundreds of thousands of citizens can act, of course, only through representatives as far as the drafting of the Constitution is concerned ; but in these cases the people have reserved to themselves expressly and unconditionally the initiative as well as the final decision." " Popular sovereignty is the sole basis, not only in theory but in practice, of the entire legal system of the Union, as well as of the several States." That our national government, in any branch of it, is beyond reach of the people ; or has any sort of " supremacy," except a limited measure of power granted by the supreme people, is an error. Judge Cooley says: " The want of sovereignty in the government, or in any branch thereof, follows so necessarily from the manner in which the Constitution was called into existence, that the tenth article of the Amendments was scarcely necessary to make it plain that the powers not delegated to the United States by the Constitution, nor prohibited by it to the States, are reserved to the States respectively or to the people." Of what use would be this limitation of the government, if the government itself can alone and supremely and finally judge whether itself is a trespasser ?

Judge Hitchcock says : " But with the people still remains the final word, the ultimate appeal, whenever the gravity of the occasion requires. And the self-imposed checks and delays of the Constitution are but obstacles, as James Russell Lowell has said, ' in the way of the people's whim, not of their will.' " The problem of

6

the Supreme Court is then how far it may extend its
will without being overruled by the Supreme Court of
Appeals—the people.

The difficulties in the way of a popular verdict do not
affect the principle that the " States or the people "
hold all power not distinctly delegated to their representatives—and these representatives remain, when
organically acting, only representatives. The two extreme positions are equally embarrassing. (1) That
the United States Government does in itself have final
power of deciding as to the constitutionality of its own
actions ; and (2) that each State has a final power of
deciding as to the constitutionality of the act of the
general government. No. (1) involves despotism ; a
conflict of the three departments ; and would end of
necessity in the final supremacy of one department.
No. (2) involves anarchy and State conflict and disruption. The people have organized both these expressions of public opinion, the State and the General
Government, as conveniences ; and will leave to each
its self-definitive power where it does not collide with
public conscience or public conviction. But when the
Supreme Court decides, or Congress acts, contrary to
the conviction or conscience of the people, remedy will
be sought ; sometimes in State action, sometimes in
political party action.

I think we have no better summing up of the Constitution than by Judge Charles A. Kent. " The construction by the Supreme Court is *most likely* to be
correct, since made by eminent judges after full argument. Public opinion has therefore established the
rule that such construction should be followed by Congress and the President. Ordinarily the decisions of
the Supreme Court on the most vital constitutional

questions are accepted as final. But a question may
be re-argued, and the decision changed. And when
public opinion demands, a ruling may be changed by
Constitutional amendment. The best judges are far
from being infallible. No one can study the decisions
of the Supreme Court without feeling that many of
them have arisen from the prejudices of the judges, and
would have been defeated at other periods in the history of the Constitution. Judges are often appointed
because of their eminence in political life. They are
always selected from the party in power. They do not
lose their political prejudices by their transfer to the
bench. The general character of a judge, his special
idiosyncracies, his personal jealousies even, may affect
his decision of the gravest constitutional question. The
Supreme Court *and* the people of the whole country are
the final judges of the extent of national power . . .
a right of revolution remaining to any State or section
complaining of injury not capable of redress in any
other way ; and such locality must judge for itself
whether or not the end justifies the revolution." Such
revolution would be justified by the conclusion whether
it was reconstructive or destructive. What will be
the line which will eventually divide State jurisdiction
from national no one can tell. Perhaps it will be a line
often changing. Judge Chamberlain in the same spirit
says : " No fact connected with our development is
more remarkable than the infrequency of serious conflict between the Constitutions of the States and of the
United States. No more striking evidence exists of the
civil capacity of the American people." The real reliance is therefore on what Bagehot calls " the civil instincts of our race." " The theory of our government
is opposed to the deposit of unlimited power anywhere.

Not only may no person say, but no power, or department, or function of government may say, '*L'État c'est moi.*' "

Mr. Von Holst allows that the decisions of the Supreme Court cannot be said to be questions " decided in a proper sense of the word ; because the Supreme Court can change its opinion, and has changed it, in Constitutional questions of the highest significance, after the lapse of a comparatively short time." " Moreover the Supreme Court is not supreme, but equal to the other factors of the federal government." The Supreme Court has simply to give to the people its opinion whether a law is allowed by the Constitution. " In a word, the domain of the Court is not politics, but law." It cannot even express its opinion until a contested case is brought before it. Congress may even take away its right to express an opinion. So we have the argument that the Supreme Court is final arbiter of all constitutional questions and laws arising under the Constitution, reduced to this minimum, Provided the people, acting as the President, or as Congress, or as States, do not prevent or overrule.

The Alien and Sedition Acts were flagrant violations of the spirit of the Constitution; they were usurpations of power that could not be tolerated by the people if they would retain their freedom. Honor and loyalty demanded action. Individual protest was answered with violence and imprisonment. No one might even ask to have these laws repealed. Petitions were unnoticed, or if noticed at all, with sneers. To appeal to the Supreme Court was absurd, for that was the most partisan branch of the government. It not only did not decide the Alien and Sedition Acts to be unconstitutional, it aided to enforce them. It was imperative

that the States that retained the spirit of freedom
should act. The wisest plan for action had not been
thought out. New ground was to be trodden. The
precise limits to the intervention of a single State had
not been discussed. The Virginia Resolutions were
drawn by no disloyalist ; but by the very father of the
Constitution. The intent was not to break the Union ;
but to prevent the Union from dissolving into an oligarchy. Mr. McMaster decides that Jefferson's part in
the Kentucky Resolutions " deprives him of all possible
claim to statemanship." Otherwise we might believe
the world not wholly in error as to the abilities of our
third and probably our greatest President.

Thirty years after the event, Mr. Madison wrote that
the Virginia Resolutions distinctly invited the States
to take necessary and proper measures for co-operation,
in adopting " necessary and proper measures for maintaining the authority, rights and liberties reserved to
the States respectfully, or to the people." Dec. 23,
1832, he writes : " The plural number, States, is in
every instance used, where reference is made to the
authority which presided over the government. As I
am now known to have drawn those documents, I aver
that the distinction was intentional." But Jefferson's
letters show that in 1798 the exact method of defending
State rights and preventing usurpation had not been
fully considered.

Writing June 1, 1798, he argues the case with John
Taylor, showing the object of the Resolutions to have
been the reverse of disloyalty to the Union. " Suppose
the New England States cut off, will our natures be
changed ? Immediately we shall see a Pennsylvania and
a Virginia party arise in the residuary confederacy,
and the public mind will be distracted with the same

party spirit.   If we reduce our Union to Virginia and
North Carolina, immediately the conflict will be established between the representatives of these two States,
and they will end by breaking into their simple units.
Who can say what would be the evils of scission, and
when and where they would end ?   Better keep together ; haul off from Europe as soon as we can, and
from attachment to any portions of it ; and if they
show their power just sufficient to hoop us together, it
will be the happiest situation in which we can exist.
For this is a game where principles are the stake."

Henry A. Wise commenting, says : " Madison did
not, in any sense, teach nullification of Congressional
acts.   Nullification lays down State rights as absolute
and supreme, which Madison denied.   Mr. Madison
announced as his position that each State might judge
for itself of the constitutionality of any act of Congress ;
but not without responsibility to each and all of the
other States."

One of the latest of Jefferson's letters, written to
Judge Johnson in 1823, reverts to the general subject
in these terms : " Can it be believed that under the
jealousies prevailing at the adoption of the Constitution, the States meant to surrender the authority of preserving order, or enforcing moral duties, and restraining
vice within their own territory ? . . .   The States
supposed that by their Tenth Amendment they had
secured themselves against constructive powers.   I ask
for no straining of words against the general government—nor yet against the States.   But the Chief-Justice says there must be an ultimate arbiter somewhere.
True, there must, but does that prove it is either party ?
The real arbiter is the people of the Union ; assembled
by their deputies in convention, at the call of Congress ;

or of two-thirds of the States." The position taken
may be summed up as an appeal to the people, as States,
to co-operate in nullifying malign legislation, endorsed
by the other departments of government. Neither Virginia nor Kentucky proposed that separate action which
would constitute revolution.

However the Alien and Sedition Acts must not be
taken alone in estimating the drift of the Federal party.
They were only a symptom of a deep political disease.
There was not only a readiness for despotic legislation,
in order to suppress criticism ; there was as certain a
drift toward European social and political methods. A
standing army of thirteen thousand was proposed in
time of peace. Besides Washington, who was to be in
supreme command, there were to be three major-gen-
erals. The relation of these was a matter of importance, for Washington was growing too old for active
service. The three men selected were Knox, Charles
Pinckney, and Hamilton. President Adams placed
Knox next to Washington, from his seniority in the
Revolutionary army. But Hamilton insisted on holding that rank himself ; and his friends intrigued to
override the President's decision. Washington was
persuaded to intervene in behalf of Hamilton. Pickering wrote that he had a very poor opinion of Washington's military ability ; avowing his real purpose to be
to make Hamilton commander-in-chief. Adams yielded,
but with indignation ; saying that Hamilton had no
claims whatever to high military rank, being a foreigner, " a person of comparatively low rank in the
late army, with no popularity anywhere in America ;
and of merits which were estimated as diversely as
those of Calvin." " If," he added, " I were to appoint
Hamilton second in rank I should consider it the most

difficult act of my life to justify.'' The object of
Hamilton in desiring to practically command an army
may have been simply vanity.

But it is a fact that Hamilton, together with Pickering, had on foot a secret plot with Great Britain, to
engage in a joint plundering expedition against Spanish
possessions in America. The United States were to
furnish an army ; England a navy ; to make conquest
of Florida, Louisiana, and Mexico. The two powers
were to divide the spoils. This would bring us into
collision with France as well as Spain ; while fixing us
in alliance with Great Britain.

Hamilton had not only angered Adams but awakened
his suspicions. From this time forward he and Pickering were watched by the President with untiring as-
siduity, not to say with hatred. Miranda, the London
agent, wrote to Hamilton, '' all is ready for your President to give the word.'' But the last thing Adams intended was to give the word to set Hamilton at the
head of an army, filibustering our neighbors. On the
contrary, he secretly determined that the country
should not be dragged into war. Just when the conspirators felt sure that France was forever alienated,
and that war must ensue, Adams amazed the country
by one morning notifying the Senate that he had determined to send a special commission to France to
negotiate a treaty.

The party rent was already past mending. This
action of the President, looking toward pacificatory
dealings with France, turned suppressed anger into
open and avowed hate. Adams had proved to be a
poor tool. Dogged as he was honest, he was determined to keep the country out of needless war. The
full program of Hamilton seems to have been (1) An

alliance with England ; (2) War upon South America
and Mexico ; (3) An army of fifty thousand men ; (4)
Himself to be commander-in-chief ; (5) A military
régime for the nation ; controlling the States as subordinate ; (6) Alien and Sedition laws to repress
criticism. Adams, on the contrary, had these three
points to his political creed : (1) Neutrality in European quarrels ; (2) If driven to an alliance, France is
our natural ally ; (3) Great Britain, as at present disposed, is the last power to which there should be resort
for any alliance, political or military. The Hamiltonians tried every means of circumvention in vain, and fell
back on abuse. Hamilton himself wrote " A letter from
Alexander Hamilton concerning the public conduct
of John Adams, Esq., President of the United States."
In this he spoke of Adams as " A man of disgusting
egotism ; of distempered jealousy ; and ungovernable
indiscretion "; which meant that Adams had opinions
and a will of his own. The lampoon should have
brought Hamilton condign punishment under the Sedition Act ; but that Act was not intended for Federals.

Of the French mission Adams wrote, ten years later,
that it was the most meritorious act of his life. " I desire," he said, " no other inscription on my gravestone
than, Here lies John Adams, who took upon himself
the responsibility of peace with France in the year
1800." Finding his cabinet treacherous, he finally dismissed three of its members ; taking John Marshall to
be Secretary of State in place of Pickering. Near the
close of his administration in 1799, the desperate Hamilton faction united in an appeal to Washington once
more to come from his retirement and take the helm.
The appeal reached him on his death bed.

The condition of affairs brought about by these in-

trigues is illustrated by the following passage from the
Rutland *Herald* of Vermont : " The intelligence that
comes from every quarter denotes an uncommon agitation of the public mind, by the late measures of the Federal representatives.   In several papers there are strong
intimations that it will be soon necessary to dissolve the
Federal Union, and not be embarrassed any longer with
the debts and negroes of the Southern States.   Calm
and prudent counsels are certainly best in the present
emergency.   And it cannot be too often inculcated
upon our citizens that their duty and safety requires
that they pay a steady regard to civil and moral considerations in every movement they make.   If it was
ever necessary to look out for calm, prudent, and judicious men for Federal representatives, now is the time.
A few hot, rash, and hasty men in Congress, and the
Federal Union will most probably be rent asunder."

Secession or dissolution was in the air. Few believed
it possible to hold together. The Alien and Sedition
Acts were indiscreet in the extreme, apart from their
tyrannical character.   Resistance became a virtue.
New England was hardly behind Virginia in its disgust.   The Middle States were turned over to Republican opposition.   John Adams wrote that all the
Southern and Western States were in unison with Virginia and Kentucky; and menaced separation.

The Resolutions of Virginia and Kentucky were in
fact an expression of national sentiment. Their publication was not a matter that went by itself. From
every corner of the land remonstrances and resolutions
were sent to the seat of government. Great rolls of
petitions, bearing tens of thousands of names, were
carried for the first time in American history, into the
Congressional halls. In Virginia and Kentucky the

people acted as States ; elsewhere as individuals or as
towns. Eighteen thousand names went up from Pennsylvania. This way of voting was at least significant.
In the spirit of the people of the whole Union, the General Assembly of Pennsylvania protested against the
palpable and alarming infractions of the Constitution,
which " by uniting legislative and judicial powers to
the executive, subvert the general principles of free
government." If this was not nullification in form, it
meant the people would not submit to the Alien and
Sedition Acts. Had the party in power been wise, it
would have treated such petitions and protests with
great consideration. It was compelled to act on repeal
resolutions ; but haughtily voted them down. It was
high time for a total change of procedure.

APPENDIX TO CHAPTER II

THE ALIEN ACT, JUNE 25, 1798

Section 1. Be it enacted by the House of Representatives of the United States of America in Congress assembled, That it be lawful for the President of the United States at any time during the continuance of this act, to order all such aliens as he may judge dangerous to the peace and safety of the United States, or shall have reasonable grounds to suspect are concerned in any treasonable or secret machinations against the government thereof, to depart out of the territory of the United States, within such time as shall be expressed in such order, which order shall be served on such alien by delivering him a copy thereof, or leaving the same at his usual abode, and returned to the office of Secretary of State, by the marshal or other person to whom the same shall be directed. And in case any alien so ordered to depart, shall be found at large within the United States after the time limited in such order for his departure, and not having obtained a license from the President to reside therein ; or having obtained such license shall not have conformed thereto, every such alien shall, on conviction thereof, be imprisoned for a term not exceeding three years, and shall never after be admitted to become a citizen of the United States. Provided always and be it further enacted, that if any alien so ordered shall prove to the satisfaction of the President, by evidence to be taken before such person or persons as the President shall direct, who are for that purpose hereby authorized to administer oaths, that no injury or danger to the United States will arise from suffering such alien to reside therein, the President may grant a license to such alien to remain within the United States for such time as he may judge proper, and at such place as he may designate. And the President may also require of such alien to enter into a bond with the United States in such penal sum as he may direct, with one or more sufficient sureties to the satisfaction of the person authorized by the President to take the same, conditioned for the good behavior of such alien during his residence in the United States, and not violating his license, which license the President may revoke whenever he shall think proper.

Sec. 2. And be it further enacted, That it shall be lawful for the President of the United States, whenever he may deem it necessary for the public safety, to order to be removed out of the territory thereof, any alien who may or shall be in prison in pursuance of this act ; and to cause to be arrested and sent out of the United States such of those aliens as shall have been ordered to depart therefrom, and shall not have obtained a license aforesaid, in all cases where, in the opinion of the President, the public safety requires a speedy removal. And if any such alien, so removed or sent out of the United States by the President, shall voluntarily return thereto, unless by permission of the President of the United States, such alien on conviction thereof, shall be imprisoned so long as, in the opinion of the President, the public safety may require.

Sec. 3. And be it further enacted, That every master or commander of any ship, which shall come into any port of the United States after the first day of July next, shall immediately on his arrival make report in writing to the collector, or other chief officer of the customs of such port, of all aliens, if any, on board his vessel ; specifying their names, age, the place of nativity, the country from which they shall have come, the nation to which they belong and owe allegiance, their occupation and a description of their persons, as far as he shall be informed thereof, and on failure, every such master and commander shall forfeit and pay three hundred dollars ; for the payment whereof on default of such master or commander, such vessel shall also be holden, and may by such collector or other officer of the customs be detained.   And it shall be the duty of such collector, or other officer of the customs, forthwith to transmit to the office of the department of State true copies of all such returns.

Sec. 4.   And be it further enacted, That the circuit and district courts of the United States, shall respectively have cognizance of all crimes and offenses against this act.   And all marshals and other officers of the United States are required to execute all precepts and orders of the President of the United States, issued in pursuance or by virtue of this act.

Sec. 5.   And be it further enacted, That it shall be lawful for any alien who may be ordered to be removed from the United States, by virtue of this act, to take with him such part of his goods, chattels, or other property, as he may find convenient ; and all property left in the United States by any alien, who may be removed as aforesaid, shall be, and remain subject to his order and disposal, in the same manner as if this act had not been passed.

Sec. 6.   And be it further enacted, That this act shall continue and be in force for and during the term of two years from the passing thereof.

Approved, June 25, 1798.

Section 1. Be it enacted by the Senate and House of Representatives of America, in Congress assembled, That if any persons shall unlawfully combine or conspire together, with intent to oppose any measure or measures of the Government of the United States, which are or shall be directed by proper authority, or to impede the operation of any law of the United States, or to intimidate or prevent any person holding a place of office in or under the government of the United States, from undertaking, performing or executing, his trust or duty ; and if any person or persons, with intent as aforesaid, shall counsel, or advise, or attempt to procure any insurrection, riot, unlawful assembly, or combination ; whether such conspiracy, threatening, counsel, or advice, or attempt shall have the proposed effect or not, he or they shall be deemed guilty of high misdemeanor, and on conviction, before any court of the United States having jurisdiction thereof, shall be punished by a fine not exceeding five thousand dollars, and by imprisonment during a term not less than six months or exceeding five years ; and further at the discretion of the court may be holden to find sureties for his good behaviour in such sum, and for such time as the said court may direct.

Sec. 2. And be it further enacted, That if any person shall write, print, utter or publish, or shall cause or procure to be written, printed, uttered or published, or shall knowingly and willingly assist or aid in writing, printing, uttering or publishing any false, scandalous, and malicious writing or writings against the government of the United States, or either house of the Congress of the United States, or the President of the United States, with intent to defame the said government, or either house of the said Congress, or the said President, or bring them, or either of them, into contempt or disrepute ; or excite against them, or either or any of them, the hatred of the good people of the United States, or to stir up sedition in the United States, or to excite any unlawful combinations therein, for opposing or resisting any law of the United States, or any act of the President of the United States, done in pursuance of any such law, or of the powers vested in him by the Constitution of the United States, or to resist, oppose, or defeat any such law or act, or to aid, encourage or abet any hostile designs of any foreign nation against the United States, their people or government, then such person, being convicted before any court of the United States having jurisdiction thereof, shall be punished by a fine not exceeding two thousand dollars, and by imprisonment not exceeding two years.

Sec. 3.   And be it further enacted and declared, That if any person be prosecuted under this act, for writing or publishing any libel aforesaid, it shall be lawful for the defendant, upon trial of the cause, to give in evidence in his defence, the truth of the matter contained in the publication charged as a libel.   And the jury who shall try the cause, shall have the right to determine the law and the fact, under the direction of the court, as in other cases.

Sec. 4.   And be it further enacted, That this act shall continue and be in force until the third day of March, one thousand eight hundred and one, and no longer : Provided, that the expiration of the act shall not prevent or defeat a prosecution and punishment of any offence against the law, during the time it shall be in force.

Approved July 14, 1798.

THE VIRGINIA RESOLUTIONS

Resolved that the General Assembly of Virginia doth unequivocally express a firm resolution to maintain and defend the Constitution of the United States, and the Constitution of this State against every aggression, either foreign or domestic, and that they will support the government of the United States in all measures warranted by the former.

That this assembly most solemnly declares a warm attachment to the union of the States, to maintain which, it pledges all its powers ; and that for this end it is their duty to watch over and oppose every infraction of those principles which constitute the only basis of that union ; because a faithful observance of them can alone secure its existence and the public happiness.

That this assembly doth explicitly and peremptorily declare that it views the powers of the Federal Government, as resulting from the compact, to which the States are parties, as limited by the plain sense and intention of the instrument constituting that compact ; as no farther valid than they are authorized by the grants enumerated in that compact ; and in case of a deliberate, palpable, and dangerous exercise of other powers not granted in that compact, the States who are parties thereto have the right, and are in duty bound to interpose, for arresting the progress of the evil, and for maintaining within their respective limits, the authorities, rights, and liberties appertaining to them.

That the General Assembly doth also express its deep regret that a spirit has, in sundry instances, been manifested by the Federal Government, to enlarge its powers, by forced constructions of the constitutional charter which defines them ; and that indications have

7

appeared of a design to expound certain general
phrases (which having been copied from the very
limited grant of powers in the former articles of confederation, were the less liable to be misconstrued), so
as to destroy the meaning and effect of the particular
enumeration, which necessarily explains and limits the
general phrases ; and so as to consolidate the States by
degrees into one sovereignty, the obvious tendency and
inevitable consequence of which would be to transform
the present republican system of the United States into
an absolute, or at best mixed monarchy.

That the General Assembly doth particularly protest
against the palpable and alarming infractions of the Constitution, in the two late cases of the "Alien and Sedition
Acts," passed at the last session of Congress, the first of
which exercises a power nowhere delegated to the Federal Government ; and which by uniting legislative and
judicial powers to those of executive, subverts the general principles of free government, as well as the particular organization and positive provisions of the federal
constitution : and the other of which acts, exercises in
a like manner a power not delegated by the constitution, but on the contrary expressly and positively forbidden by one of the amendments thereto ; a power
which more than any other ought to produce universal
alarm, because it is levelled against the right of freely
examining public characters and measures, and of free
communication among the people thereon, which has
ever been justly deemed the only effectual guardian of
every other right.

That this State having by its convention which ratified the Federal Constitution, expressly declared, " that
among other essential rights, the liberty of conscience
and of the press cannot be cancelled, abridged, re-

strained, or modified by any authority of the United States," and from its extreme anxiety to guard these rights from every possible attack of sophistry or ambition, having, with other States, recommended an amendment for that purpose, which amendment was in due time annexed to the Constitution, it would mark a reproachful inconsistency and criminal degeneracy, if an indifference were now shown to the most palpable violation of one of the rights thus declared and secured, and to the establishment of a precedent which may be fatal to the other.

That the good people of this commonwealth, having ever felt and continuing to feel the most sincere affection to their brethren of the other States, the truest anxiety for establishing and perpetuating the union of all, and the most scrupulous fidelity to that constitution which is the pledge of mutual friendship, and the instrument of mutual happiness ; The General Assembly doth solemnly appeal to the like dispositions of the other States, in confidence that they will concur with this commonwealth in declaring, as it does hereby declare, that the acts aforesaid are unconstitutional, and that the necessary and proper measures will be taken by each for co-operating with this State, in maintaining unimpaired the authorities, rights and liberties reserved to the States respectively, or to the people.

That the Governor be desired to transmit a copy of the foregoing resolutions to the executive authority of each of the other States, with a request, that the same may be communicated to the legislature thereof.

And that a copy be furnished to each of the Senators and Representatives, representing this State in the Congress of the United States.

December the 24th 1798.

MR. MADISON'S REPORT ON THE VIRGINIA
RESOLUTIONS

House of Delegates, Session of 1799–1800.

. . . The resolution declares ; first, that " it views
the powers of the Federal Government, as resulting
from the compact to which the States are parties," in
other words, that the Federal powers are derived from
the Constitution ; and that the Constitution is a compact, to which the States are parties.

Clear as the position must seem, that the Federal
powers are derived from the Constitution, and from
that alone, the committee are not unapprized of a late
doctrine, which opens another source of Federal powers,
not less extensive and important, than it is new and
unexpected. The examination of this doctrine will be
most conveniently connected with a review of a succeeding resolution. The committee satisfy themselves here
with briefly remarking, that in all the contemporary
discussions and comments which the Constitution under-
went, it was constantly justified and recommended, on
the ground that powers not given to the Government,
were withheld from it ; and, that if any doubt could
have existed on this subject, under the orginal text of
the Constitution, it is removed, as far as words could
remove it, by the 12th amendment, now a part of the
Constitution, which expressly declares, " that the
powers not delegated to the United States, by the Constitution, nor prohibited by it to the States, are reserved to the States respectively or to the people."

The other position involved in this branch of the
resolution, namely, that " the States are parties to the
Constitution or compact," is, in the judgment of the
committee, equally free from objection. It is indeed

true that the term "States" is sometimes used in a
vague sense, and sometimes in different senses, according to the subject to which it is applied. Thus, it
sometimes means the separate sections of territory occupied by the political societies within each : sometimes
the particular governments, established by those societies ; sometimes those societies as organized into those
particular governments ; and, lastly, it means the people composing those political societies, in their highest
sovereign capacity. Although it might be wished that
the perfection of language admitted less diversity in the
signification of the same words, yet little inconvenience
is produced by it, where the true sense can be collected
with certainty from the different applications. In the
present instance, whatever different construction of the
term "States," in the resolution may have been entertained, all will at least concur in the last mentioned ;
because in that sense, the constitution was submitted to
the "States ;" in that sense the "States" ratified it :
and in that sense of the term "States" they are consequently parties to the compact from which the powers
of the Federal Government result.

The resolution having taken this view of the Federal
compact, proceeds to infer, "That in case of a deliberate, palpable, and dangerous exercise of other powers,
not granted by the said compact, the States who are
parties thereto, have the right, and are duty bound to
interpose for arresting the progress of the evil, and for
maintaining within their respective limits, the authorities, rights, and liberties appertaining to them."

It appears, to your committee to be a plain principle,
founded in common sense, illustrated by common practice, and essential to the nature of compacts—that
where resort can be had to no tribunal superior to the

authority of the parties, the parties themselves must be the rightful judges in the last resort, whether the bargain made has been pursued or violated. The Constitution of the United States was framed by the sanction of the States, given by each in its sovereign capacity. It adds to the stability and dignity, as well as to the authority of the Constitution, that it rests on this legitimate and solid foundation. The States then, being the parties to the constitutional compact, and in their sovereign capacity, it follows of necessity, that there can be no tribunal above their authority, to decide in the last resort, whether the compact made by them be violated ; and consequently, that, as the parties to it, they must themselves decide in the last resort, such questions as may be of sufficient magnitude to require their interposition.

The resolution has, accordingly, guarded against any misapprehension of its object, by expressly requiring for such an interposition, " the case of a deliberate, palpable, and dangerous breach of the Constitution, by the exercise of powers not granted by it." It must be a case not of a light and transient nature, but of a nature dangerous to the great purposes for which the Constitution was established. It must be a case, moreover, not obscure or doubtful in its construction, but plain and palpable. Lastly, it must be a case not resulting from a partial consideration, or hasty determination ; but a case stampt with a final consideration and deliberate adherence. It is not necessary, because the resolution does not require, that the question should be discussed, how far the exercise of any particular power, ungranted by the Constitution, would justify the interposition of the parties to it. As cases might easily be stated, which none would contend ought to fall within that description—cases, on the other hand, might with equal ease, be stated so flagrant and so fatal, as to unite every opinion in placing them within the description.

But it is objected that the Judicial authority is to be regarded as the sole expositor of the Constitution in the last resort ; and it may be asked for what reason, the declaration by the General Assembly, supposing it to be theoretically true, could be required at the present day, and in so solemn a manner.

On this objection it might be observed: first, that there may be instances of usurped power, which the forms of the Constitution would never draw within the control of the Judicial department ; secondly, that if the decision of the Judiciary be raised above the authority of the sovereign parties to the Constitution, the decisions of the other departments not carried by the forms of the Constitution before the Judiciary, must be equally authoritative and final with the decisions of that department.   But the proper answer to the objection is, that the resolution of the General Assembly relates to those great and extraordinary cases, in which all the forms of the Constitution may prove ineffectual against infractions dangerous to the essential rights of the parties to it.   The resolution supposes that dangerous powers not delegated, may not only be usurped and executed by the other departments, but that the Judicial department, also, may exercise or sanction dangerous powers beyond the grant of the Constitution ; and, consequently, that the ultimate right of the parties to the Constitution, to judge whether the compact has been dangerously violated, must extend to violations by one delegated authority, as well as another ; by the Judiciary, as well as the Executive, or the Legislative.

However true, therefore, it may be that the Judicial departments, in all questions submitted to it by the forms of the Constitution, to decide in the last resort, this resort must necessarily be deemed the last in relation to the authorities of the other departments of the government ; not in relation to the rights of the parties to the constitutional compact ; from which the Judicial as well as the other departments hold their delegated trusts. On any other hypothesis, the delegation of Judicial power would annul the authority delegating it ; and the concurrence of this department with others in usurped powers, might subvert forever, and beyond the possible reach of any rightful remedy, the very Constitution, which all were instituted to preserve.

The truth declared in the resolution being established, the expediency of making the declaration at the present day, may safely be left to the temperate consideration and candid judgment of the American public. It will be remembered, that a frequent recurrence to fundamental principles, is solemnly enjoined by most of the State Constitutions, and particularly by our own, as a necessary safeguard against the danger of degeneracy to which republics are liable, as well as other governments ; though in a less degree than others. And a fair comparison of the political doctrines not unfrequent at the present day, with those which characterized the epoch of our Revolution, and which form the basis of our Republican Constitutions, will best determine whether the declaratory recurrence here made to those principles ought to be viewed as unreasonable and improper, or as a vigilant discharge of an important duty. The authority of Constitutions over Governments, and of the sovereignty of the people over Constitutions are truths, which are at all times necessary to be kept in mind ; and at no time, perhaps, more necessary than at present.

CHAPTER III

WHILE the fundamental distinction of the two parties, Federal and Republican, lay in the very nature of things, there was really but one party organization for the first twelve years of the national life. The Republican name was slowly adopted to cover a vastly incoherent opposition. As soon as the drift of those in power to construe the Constitution to suit their centralizing if not monarchical views, was perceived, opposition became organic.

Hamilton wrote soon after the Constitutional Convention : " It may triumph altogether over the State governments, and reduce them to entire subordination, dividing the large States into simpler districts. The *organs* of the general government may also acquire additional strength." Madison wrote that he separated from Hamilton because the latter wished " to administer the government into what he thought it ought to be ; while on my part I endeavored to make it conform to the Constitution as understood by the Convention that produced it ; and particularly by the State conventions that adopted it."

Opposition was for some time rather to measures

than to the Federal party. This antagonism was often
passionate, occasionally unreasonable and mobocratic.
There was a drift to bring about a necessary choice
between anarchy and despotism. The disaffection
shown so widely toward the Jay Treaty was a spon-
taneity of popular sentiment ; an outburst not of party
antagonism but of popular disgust. To some extent
the nullification resolutions of 1798 indicated a stage
over to more general political measures and method—
tentative but resolute. The country stood in need of a
second party to give orderly voice to opinion and sentiment.

The Federalists had been accused of being Anglomen.
This love for England was, however, the only reason
why the party had coherence at all. It was the strong
English instinct in the American people that brought
them into any degree of co-operation—that taught
them by heredity how to put a governmental machine
in operation. New England especially had English sentiment. Decidedly independent, made up of the outs
of English political and church life ; it was, however,
far from anarchical in temperament. Its Parliamentary history up to the Revolution had been peculiarly tactful. Massachusetts had governed herself for
the first fifty years of her existence—she had lived
under two charters and had made two constitutions.
Entering into a general Union, New England was partial to strong government ; but was equally fond of
governing. So it came about that while she gave the
Union its cradle, she was not ready to let it out of its
cradle.

Although Anglo-Saxon instincts and institutions
created a great and common foundation for government
in both countries, the breach between the United States

and Great Britain was more vital than was assumed. The union of the States was established on commercial freedom. It also recognized not only that all government is from and of the people, but that it remains with the people. Government by "The Best," although selected by the people, proved as serious a failure as government by a single monarch chosen *dei gratia*. This was the lesson taught by the Federalist party. There is no question as to the real quality of these men. They were the best—the ablest men in America. But aristocracy has ever constituted the worst government in the world. It is by its nature a Ring ; a Ring is an aristocracy. Invariably the people have fled from aristocracies to monarchy. Freytag shows that absolutism, of necessity, develops a lack of common sense, ending in mania or idiocy. It will not be surprising if this sort of insanity sometimes manifests itself under republican forms of government. .

With a growing debt behind, with discord at home and abroad, with new problems of vital importance ahead, the Federal party found itself broken into three fragments. Adams with a large section veered toward a broader nationalism ; Pickering, Griswold, Cabot, Wolcott, and Strong held to the course marked out by the Alien and Sedition laws ; Hamilton agreed with neither section ; and while valued for his genius, was not trusted by any.

The young nation, as it approached 1800, was about to transfer its interests from those who believed they alone were fit to govern, to a new party. The "Republicans" were henceforth to be held responsible for the sacred cause of liberty. The Federal party indeed was dead ; although Federalism was not dead. It was never so much alive as after it ceased to be the governing force. Its existence will be attested as we shall see by several episodes of a sort we should be glad to expunge from our history.

While the two parties are in their death grapples in 1800, it will be worth our while to compare them as they then stand. The old party, now doomed to defeat, owed much of its trouble to that unfortunate element that had remained loyal to Parliament during the Revolution. Robert Livingston in January, 1784, wrote, " Our parties are, first the Tories who still hope for power. Second, the violent Whigs who are for expelling all Tories from the State ; and third those who wish to suppress violence, and soften the rigor of the laws against loyalists." The Whigs issued an address, warning the people against attempts that would be made, imperceptibly to change the spirit of the government. Hamilton championed the moderates, and soon the Tories had political equality. This they exercised by throwing all their votes and influence for the Federalists, who grew out of the moderates.

It is equally important to note that the Federalist party from first to last was a party of leaders, and never a party of the people. It was based on a distrust of the people. " A Democracy," said Dennis' Portfolio, " is scarcely tolerable at any period. It is on trial here ; and the issue will be civil war, desolation, and anarchy." Fisher Ames said, " Our country is too big for Union, too sordid for patriotism, too democratic for liberty. Its vices will govern it by practicing upon its follies." Cabot said, " I hold democracy to be the government of the worst." Adams was by instinct democratic, by education autocratic ; but he was the truest to the people of all the Federal leaders. At times, however, he was very impatient and impetuous. In 1798 he said : " As to trusting to a popular assembly for the preservation of our liberties, it is the merest chimera. They never had any rule but their own will—and I would as lief be again in the hands of our old committees of safety who made the law and`executed it at the same time." Cabot in 1793 is credibly reported to have openly advocated à President for life and a hereditary Senate. Hildreth thinks the Federalists " never " were a majority of the people ; and hardly for a single session were they in the majority in Congress.

After the withdrawal into opposition of Madison and Jefferson, there were pre-eminent as leaders, Washington, Adams, and Hamilton. Washington seems never to have been quite assured of the perpetuity of the republican form of government. But inflexibly determined to give it a fair trial, he bent his energies to create a national sentiment. A favorite measure which he constantly urged on Congress was a National University. " It has been," he wrote, " my ardent wish to see a plan devised to spread systematic ideas throughout all parts of this rising empire—thereby doing away with local attachments and State prejudices." Adams has been charged with jealousy of Washington. It is certain that he never felt that Washington was his superior as a statesman. But while lacking the superb poise of character of Washington, and possessing less administrative ability, he was the peer of any American of his day in intellectual power. Before the close of his administration he was well on his way out of Federalism. His associations were with one party ; his convictions with the other.

Hamilton must be counted by himself in this estimate. He was not a Federalist ; nor was he a New Englander in sentiment. He was not in more harmony with

Adams than with Clinton, or with Jefferson. He was a prototype of the Independent in Politics. His classification with the Federals is only because he believed in a strong government. This brought him in closer affiliation with Washington than with Jefferson. But his course was utterly disorganizing, and inside any party there could be for him no rival tolerated. So it has come about that Hamilton has secured a unique place in our political history. Not one of his measures was adopted except the financial; and we are now, at the end of a century, compelled to find out if this has not been our undoing. But his brain so teemed with expedients, that the unseated party which had rejected his policies, and refused to nominate him for the Presidency, began to claim him as the greatest, wisest, most wonderful genius of Federalism.

During the last four years of Federal rule Hamilton had the advantage of being the critic, while Adams bore the responsibility of government. When the party found itself deserted by the people it was easy to believe that the fault was with the President. It is easier for us now to see that had Hamilton been faithful to Adams the party would not have been so summarily dismissed. Almost the autocrat of Washington's administration, he had found Adams by no means as confident of his genius or his honesty. Indeed Adams stood less in need of Hamilton than had Washington. The latter never felt quite sure of his footing until he had submitted his slowly evolved opinions to the ardent flash-light illumination of Hamilton's intellect. Washington brought out the best of the fiery West-Indian; Adams stirred him to do his worst.

A party made up of leaders and not of the people was doomed to dissolution. The final collapse found Hamilton and Adams each in command of a fragment of the party—while such lesser lights as Pickering, Ames, Wolcott, Strong and Griswold were ready for conspiracies as opportunity afforded.

The sharp contrast between the leaders of the Federal party and those of the Republicans was that the former insisted on strict construction of the Constitution ; while the latter would construe it for party or immediate ends. With rigid adherence to the letter of the Constitution, the Republicans avowed and felt a confidence in the people to govern themselves. To such men there was no treason like that to the people. The avowal of monarchical tastes or sentiments was to go back of the Revolution, and waste all of their hard-won history. The Federal party believed in government more than in the people ; the Republican party held first to the people and less to government. It was perfectly in line with Republicanism for Jefferson to say that he held an insurrection once in twenty years a wholesome feature of national life.

But the Republican party above all consisted of " the people " themselves—the great mass of common folk who constituted the body of the nation. These had voted for Washington, and mainly for Adams ; but scarcely could have been classed as Federals or Anti-federals. Anti-federals did exist before the Constitution was adopted ; but after 1790, and Rhode Island had finally accepted the Constitution, no Anti-federalism ever manifested itself. The opponents of the Federal party were other people. Jefferson in 1789 wrote : " I am not a Federalist, because I never subscribe the whole system of my opinions to the creed of any party of men whatever, in religion, philosophy, in politics or in anything else, where I am capable of thinking for myself. If I could not go to heaven but with a party I would not go at all. Therefore I protest to you I am not of the party of Federals. But I am much farther from that of the Anti-federalists." This was a very general fact. The bulk of the people never belonged to the Federalists, except as friends to the Constitution. The Constitution adopted, they were in a measure independents; but continued for a while to vote for the organization that shrewdly retained the Federal cognomen. The coalescence of a Republican party was vastly accelerated by the sentiment of gratitude to the French for their aid during our Revolution. It seemed to many that this aid called for a hearty response of sympathy during the trials of the revolution that followed in France. Marshall, in his *Life of Washington* gives us a very complete illustration of this spirit. He records the following in a list of toasts responded to in Philadelphia, at a meeting of the officers of the 2d Regiment on the 14th day of July, the first anniversary of the destruction of the Bastille :

" The fourteenth day of July, may it be a Sabbath in the calendar of freedom, and a jubilee to the European world.

Nerve to the arm, fortitude to the heart, and triumph to the soul struggling for the rights of man.

May the sister republics of France and America be as incorporate as light and heat, and the man who endeavors to disunite them be viewed as the Arnold of his country.

May honor and probity be the principles by which the connections of free nations shall be determined, and no Machiavellian commentaries explain the texts of treaties.

The treaty of alliance with France ; may those who

attempt to evade or violate the political obligation. and faith of our country be considered as traitors, and consigned to infamy.

The republics of France and America ; may the cause of liberty ever be a bond of union between the two nations.

A dagger to the bosom of that man who makes patriotism a cover to his ambition, and steals his country's happiness, absorbed in his own.

Union and mutual confidence to the patriots of France ; confusion and distress to the councils of their enemies.

May the succeeding generation wonder that such beings as kings were ever permitted to exist.

The rule of proportion; as France acted with respect to America, so may America act with respect to France.''

It is equally illustrative of Federal sentiment that Mr. Marshall should pronounce these to be examples of '' excessive and passionate devotion to France, and of equal hostility to the party in power.'' Clearly our government had drifted so strongly into Anglomania that an expression of gratitude to the French, and of general devotion to republican principles everywhere, was conceived by the Federal party to be absurd as well as dangerous sympathy for popular government.

The following toasts, which were given at a dinner in Philadelphia to express the public sentiment concerning the Jay treaty, were equally offensive to the party in power :

'' The republic of France, whose triumphs have made this day a jubilee ; may she destroy the race of kings, and may their broken sceptres and crowns, like the bones and teeth of the mammoth, be the only evidence that such monsters ever infested the earth.

8

The republic of France ; may all free nations learn of her to transfer their attachments from men to principles and from individuals to the people.

The Republic of Holland ; may her two sisters the republics of France and America form with her an invincible triumvirate in the cause of liberty.

The republic of Holland ; may that government that they are about establishing have neither the balance of aristocracy nor the checks of monarchy.

The republic of America ; may the sentiment that impelled her to resist a British tyrant's will, and the energy which rendered it effectual, prompt her to repel usurpation in whatever shape it may assail her.

The republic of America ; may the alliance formed between her and France acquire vigor with age ; and that man be branded as the enemy of liberty who shall endeavor to weaken or unhinge it."

These were not wild, or passionate, or violent expressions of factionalism, but the natural language of a youthful nation, full of the enthusiasm of a few experiences. On the other hand the language addressed to Washington was often richly aglow with similar sentiments. When he went to New York to be inaugurated, his neighbors bade him God speed saying, " Farewell ! and make a grateful people happy ; and may the Being who maketh and unmaketh at His will restore to us again the best of men and the most beloved fellow-citizen." Everywhere matrons crowded the highways, flinging flowers ; and girls dressed in white were singing odes in his honor. Birthday songs told Washington that

> " Arrayed in glory bright
> Columbia's savior comes."

It is very easy to look back and laugh at an enthusiasm. But the fact was the American people *did* sweep out nine-tenths of aristocracy—much of it as harmless as the stuff retained. America was not injured by its generous gratitude toward France. Cockades and liberty poles did no harm, nor would any harm have followed the abolition of such trash as " His Honor the Mayor," and many other scraps of titular glory. Honors in church and in state in America have proved as childish as they are empty. Marshall, however, wishes us to believe that, " In every part of the United States the love of France appeared to be a passion much more active with immense numbers than that of America." In fact the Federal party had identified fondness for itself with loyalty to America.

The existence of the Republican party was identified, and justly so, with a measure of religious independence. This brought together, at first with intenser partisanship, New England on the one hand, and Virginia with Pennsylvania and the newer territories on the other. Fanaticism added to party rancor. But meanwhile a spirit of religious freedom was slowly growing in Massachusetts and Connecticut ; while Vermont and Western New York and Ohio, which were children of these States, had left behind that entire loyalty to creed which had characterized the mother States. It was not without much of this leavening that these States were led over to the Republican ranks, very soon after its organization. Virginia had imported primogeniture and an established church. Both these it abrogated at the instigation of Madison and Jefferson. The middle and southern States began to identify religious with civil liberty ; while the leaders were fairly classed as free thinkers. Jefferson was especially obnoxious to

the Calvinists.   There seems to have been a substantial
accord between Franklin, Jefferson, Washington, and
Madison ; they were all free-thinkers.   In his later
years John Adams was fully of the same mind.   Jefferson says of his own views that " They ought to dis-
please neither the rational Christian nor Deist ; and
would reconcile many to a character they have too
hastily rejected.   I do not know that it would reconcile
the *genus irritabile vatum*, who are all in arms against
me.   They believe that any portion of power confided
to me will be exerted in opposition to their schemes.
And they believe rightly ; for I have sworn upon the
altar of God, eternal hostility against every form of
tyranny over the mind of man."

Thomas Jefferson was recognized as the leader of
the new force in American politics.   There was no one
who was considered his rival.   As Washington had
stood alone at the head of Federalism, Jefferson stood out
alone at the head of Republicanism.   Almost dreading
public life, he was gifted to shape the destinies of the
people as no other man ever did, or ever after could.

Receiving in 1797 68 electoral votes to Adams 71,
Jefferson acted as Vice-President for the next four
years.   In 1800 he was nominated for the Presidency,
with Burr on the same ticket, while Adams stood as the
general candidate of the Federals, and Pinckney was
sustained as the candidate of the Hamiltonians.   The
struggle to prevent the election of Jefferson and the
triumph of the new party was desperate.   Hamilton,
in response to the toast, " A Strong Government,"
said, " If Mr. Pinckney is not elected, a revolution will
be the consequence ; and within four years I will lose
my head, or be the leader of a triumphant army."   He
was, after Washington, the Major-General of an army,

proposed by the Federals, but that the Republicans
soon took care should not materialize. The Federal
clergy joined furiously in the opposition to Jefferson as
an atheist. While the Sedition law quickly jailed and
fined indiscreet Republicans, it allowed Federalists to
say whatever they chose. The campaign brought out
the fact that Mr. Jefferson was the best loved and the
most soundly hated man in America.

An equal number of electoral ballots having been,
from lack of party organization, cast for Jefferson and
Burr, the election was thrown into the House of Representatives. The Federalists crowned their shame by
a desperate resolve to drop their own candidates in the
House, and circumvent the people by turning the
country over to Burr, a man they themselves abhorred.
If there had before been some confidence to be placed
in their honesty, and capacity for conducting a republican government, it was now lost. Seventeen ballots
were taken in the House on Wednesday the eleventh
of February. Eight States supported Jefferson and six
Burr. There were sixteen States; Jefferson must secure nine. Vermont and Maryland were divided. An
all-night's session followed; but the struggle was not
ended. The House remained in session nominally
without adjournment for a full week, constantly balloting without effect. Burr meanwhile was more than
willing to take what he well knew the people never intended for him. Hamilton, who had labored by even
disreputable methods, to prevent the election of Jefferson, now cast all his weight against Burr, as between
the two. "But," he wrote to Wolcott, "it will be
well enough to throw out a lure to tempt Burr, and so
lay the foundation for dissension between the two
chiefs." He could however have little power to influ-

ence voters, for it was well known that in New York
he and Burr were keen rivals for political influence.
Adams and his friends had good reason for suspecting
Hamilton's motives, for they had suffered bitterly from
his pen and his tongue.   Of Burr's chance, Adams
wrote with indignation that "all the old patriots, all
the long experience of both Federalists and their opponents should be subjected to the humiliation of seeing
this dextrous gentleman rise over them."   However
on the thirty-sixth ballot the deadlock was broken.
Bayard threw a blank for Delaware ; Morris left Vermont to the Republican delegate ; and the Maryland
Federalists withdrew.   Jefferson was made President
of the United States ; and Burr was made Vice-President.   The will of the people triumphed.   During the
balloting the whole country was in a ferment.   Threats
of all sorts were rife.   It is not improbable that any
other result of the balloting would have caused a revolution.   Still we find Jefferson himself writing to Governor McKean : " Had it terminated in the elevation of
Mr. Burr, every Republican would, I am sure, have
acquiesced in a moment ; because, however it might
have been variant from the intentions of the voters, it
would have been agreeable to the Constitution.   No
man would have more cheerfully submitted than myself."   It is not certain that the people would have
submitted as readily as their leader.

Jefferson declared he would recognize no party.
" We are," he said, in his inaugural, " all Republicans, and we are all Federals.   I know that some fear
that a republican government cannot be strong—that
this government is not strong enough.   I believe this
on the contrary the strongest government on earth.   I
believe it the only one where every man, at the call of

law, would fly to the standard of the law, and would meet invasion of the public order as his own private concern." While the plot was working in Congress, he wrote to Madison that, as soon as the election was settled, he proposed " to aim at a candid understanding with Mr. Adams. I hope to induce in him dispositions liberal and accommodating." His inaugural followed out this policy of aiming to unite all honest men, for an honest government. Adams unfortunately was in a temper, and did not get over it before he had done what his own honor in after life disapproved. He filled every office with Federals ; signing commissions till the evening he must vacate the President's mansion.

Jefferson wrote to John Dickinson : " I hope to see shortly a perfect consolidation ; to effect which nothing shall be spared on my part, short of the abandonment of the principle of revolution. A just and solid republican government maintained here, will be a standing monument and example for the aim and imitation of the people of other countries ; and I join with you in the hope and belief that they will see, from our example, that a free government is of all others the most energetic ; that the inquiry which has been excited among the mass of mankind by our revolution and its consequences, will ameliorate the condition of man over a great portion of the globe." It was his passion to unite the people in one friendly alliance for good government.

He wrote to Robert Livingston : " The Constitution to which we all are attached was meant to be republican ; and we believe it to be republican, according to every candid interpretation. Yet we have seen it so interpreted and administered as to be truly, as the French say, a *monarche masque*. To put her on the

republican tack will require all the skill, firmness
and zeal of her ablest and best friends.   It is essential to assemble in the outset persons to compose our
administrative whose talents, integrity, and revolutionary fame and principles may inspire the nation
at once with unbounded confidence." To Giles he
wrote concerning offices, that it was necessary to remove some of Mr. Adams' appointees ; and he had decided to follow this rule, to remove " (1) all appointees
to civil offices, during pleasure, made after the result
of election was certainly known to Mr. Adams.   (2)
Officers who have been guilty of official misconduct.
(3) No others except in the case of attorneys and marshals." These he held it indispensable should be republican, as the courts were so decidedly Federal.
" The changes will be few, and governed by strict rule,
and not by party passion.   The right of opinion shall
suffer no invasion from me." The importance of this
decision can be realized only when we recall that
Washington had held to the idea that offices were
party property.   He wrote to Timothy Pickering,
Sept. 27, 1775: " I shall not, whilst I have the honor
to administer the government, knowingly, bring a man
into any office of consequence, whose political tenets
are adverse to the measures which the General Government is pursuing."

To Elbridge Gerry Jefferson wrote the following platform of his principles : " I do then, with sincere zeal,
wish an inviolable preservation of our present federal
Constitution, according to the true sense in which it was
adopted by the States ; that in which it was advocated
by its friends, and not that which its enemies apprehended, who therefore, became its enemies : and I am
opposed to the monarchising its features by the forms

of its administration, with a view to conciliate first a transition to a President and Senate for life, and from that to an hereditary tenure of these offices ; and thus to worm out the elective principle. I am for preserving to the States the powers not yielded by them to the Union; and to the legislature of the Union its constitutional share in the division of powers ; and I am not for transferring all the powers of the States to the General Government, and all those of that Government to the executive branch. I am for a government rigorously frugal and simple, applying all the possible savings of the public revenue to the discharge of the national debt : and not for a multiplication of officers and salaries merely to make partisans ; and for increasing, by every device, the public debt, on the principle of its being a public blessing. I am for relying, for internal defence, on our militia solely, till actual invasion ; and for such a naval force only as may protect our coasts and harbors from such depredations as we have experienced : and not for a standing army in time of peace, which may overawe the public sentiment ; nor for a navy, which, by its own expenses and the eternal wars in which it will implicate us, will grind us with public burthens, and sink us under them. I am for free commerce with all nations ; political connection with none ; and little or no diplomatic establishment. And I am not for linking ourselves, by new treaties, with the quarrels of Europe ; entering that field of slaughter to preserve their balance, or joining in the confederacy of kings to war against the principles of liberty. I am for freedom of religion, and against all maneuvres to bring about a legal ascendancy of one sect over another : for freedom of the press ; and against all violations of the Constitution, to silence by force and not by reason, the complaints or criticisms, just or unjust, of our citizens against the conduct of their agents. And I am for encouraging the progress of science in all its branches : The first object of my heart is my own country. In that is embarked my family, my fortune, and my own existence. I have not one farthing of interest, not one fibre of attachment out of it, nor a single motive of preference of any one nation to another, but in proportion as they are more or less friendly to us."

" The Republicans," said Giles, who was leader in the House, " contend that the doctrine of patronage is repugnant to the opinions and feeling of the people " ; while the Federalists, distrusting the popular sentiment, " rely on patronage, or the creation of partial interests for the support of the government." He held that the United States debt was funded, and a standing army and navy created, mainly to concentrate power. Bayard replied, asking how far their opponents intended to go in undoing the work of the Federalists. The answer was given in deeds. The Republicans did not reverse the work of their predecessors farther than to sweep away tyrannical legislation, reduce the appanage of the executive, and sharply warn the judiciary to attend to non-partisan business. Randolph declared the purpose to be " to give the death-blow to rendering the judiciary a hospital for decayed politicians," and to check the arrogance of Congress in assuming powers prohibited by the Constitution.

All of the instincts of Jefferson were for peace. He had advocated from the outset no entanglements with either of the European nations ; but peace with all. Instead of 1800 marking a victory of the French faction, it marked the death of French sentiment of a

factional sort. To Gerry he wrote : " Our countrymen
have divided themselves by such strong affections to
the French and to the English that nothing will secure
us internally but a divorce from both nations ; and this
must be the object of every real American, and its
attainment is practicable without much self-denial."
He was therefore prepared to build a party that was
not a mere faction ; and such the Republican party
eminently was. It took shape finally as the strict con-
structionist, the democratic, the economic.

There was one man, James Madison, to whom Jefferson turned, with absolute confidence in his logic, his
judgment, and his sincerity. Others, like Monroe and
Dickinson and Gerry and Edmund Randolph and Dr.
Rush, held much of his confidence ; while he was at
heart a warm lover of John Adams. Gallatin was one
of the men, detested by the Federals, whom Jefferson
brought to the front. His financial measures, precisely
the opposite of those of Hamilton, became the basis of
Republican prosperity. He lowered the taxes and he
paid the debt. Jefferson was in a marked degree like
Franklin ; for, whatever his political cares, he never
lacked deep and scientific interest in every new discovery. He imported new trees and plants ; extended
our market by private correspondence ; was able to discuss with acumen the discovery of mastodon bones, or
a problem in sociology or the arts. Most remarkable
however was his devotion to ethics in politics. His
faith in principles and faith in the people were coequal.
He believed in the Golden Rule as an international
law. His antagonism to war-establishments was based
on the conviction that if you want war you must always
appear prepared for war ; while nothing so fosters peace
as peaceful arts and peaceful methods.

Called at the threshold of the 19th century to take
the helm of the young Republic, Jefferson proved to be
pre-eminently the strongest character of that remarkable age. For breadth of purpose, for forecast, for constructive power, based on a vast ability to idealize, he
had no peer. On the other side of the Atlantic there
was one man who had equal fearlessness in breaking
with precedent ; and equal constructive ability ; but
Napoleon lived for himself primarily, for the people
afterward. Yet the brilliant intellect of both these
men had much similarity, both in method and results.
Each saw that education must be the basis of all civic
reform and national progress. Napoleon's University
scheme is still that under which the French schools are
organized; and in America we are to-day struggling to
complete the Jeffersonian system of education. He
would have the schools of each State graded from the
primaries to a State university ; and all the State universities federalized at Washington. Had we fully
heeded the insistent advice of Jefferson concerning
slavery we should not only have avoided civil war but
would have had no contest in our social institutions.
To him we owe it that involuntary servitude was debarred from the Northwest ; and only by one vote did
he fail to shut it out from the great Southwest.

Inaugurated with simplicity, Jefferson began an administration of economy, that led to the reduction of
taxation, and the steady payment of the national debt.
The effort to extend the Alien and Sedition Acts was
prevented. The Federal judiciary was compelled to
stop packing juries and making stump speeches. A
treaty with France was negotiated that was the very
opposite of the Jay treaty with Great Britain, for it
protected our rights and gave us equality among the

nations. In 1803 while negotiating with France for so much of Louisiana as should enable us to use the Mississippi, Napoleon suddenly offered to sell the whole territory. Jefferson at once closed with the offer ; and for fifteen millions made the United States owner of a domain so vast that it more than covered the original thirteen States ; how much more neither seller nor buyer could tell. The people were now almost *en masse* the followers of Jefferson ; and readily endorsed the purchase and the expense ; for said they, " while the Federals increased the national debt eight millons in five years, Jefferson has decreased it five millions in two years." State after State turned over to Republicanism ; and the whole country passed into an era of prosperity and good will.

There was at last in 1803 nothing left of the Federal party but a gang of hopelessly disaffected leaders. These men who had been recreant to their trust, perhaps never had understood Republican government, found themselves stranded without hope. Adams, who had deserted them, was a thoroughly honest man ; and when he saw the work of Jefferson to be commendable, he gave him his support, and finally his warmest friendship. Other great leaders left the party in disgust. Its history had been one of high taxation, high salaries, usurpation of power, despotic legislation, an army organized without a war, and the leaders wrangling over the command of it. Intrigue and corruption and tyranny had been the triumvirate of its short rule. With Washington it had been sheltered ; and made respectable by the mantle of his character. Under Adams it had rioted to its own disruption.

Burr bore his defeat as well as possible. But it was plain that from that hour he was politically an outcast.

Neither party trusted him ; and he was well aware that he had forfeited claim to confidence. He filled the role of Vice-President with dignity and quiet self-control. As soon as possible he proposed to retire into some other office of honor. This he suggested to Jefferson. But Jefferson frankly told him he was not to be trusted. Hamilton, Burr understood ; and he comprehended that if he ever again secured power it must be by a wrestle to the death with this brilliant politican. Both together could not rule in New York ; and hence one or the other must be master. He bided his time.

Burr's temperament was every whit as haughty as that of Hamilton. Out of the choicest stock of New England; with unequalled skill at the bar ; a fine presiding officer ; capable of drawing about him the young ; and better at combining forces than Hamilton ; he was also resolute, and capable of remarkable self-control.

Before the close of his official career as Vice-President New York was to elect a governor. Burr decided to leave Washington and enter the lists as candidate for his State's highest office. He secured the nomination of the Republicans ; but it was plain that the party was not behind him. The Federalists held a caucus with the intent of throwing their influence for Burr. But Hamilton entered the caucus determined that the man he had made his special enemy should not now be lifted over his head. His personal influence, with common sense on his side, prevailed, and Burr failed of endorsement.

Of the Federal leaders there still remained in Washington, among others, Tracy, Griswold, Plumer, and Pickering formerly of Pennsylvania, now of Massachusetts. These beheld with dismay and horror the dissolution of the party and their own loss of power.

Accustomed to rule, of a ruling caste, they now not only found themselves turned out of the offices of the nation, but Republicanism pursuing them into their own States, and depriving them of emolument and power, where before the Union they had been omnipotent. The South clearly had invaded their rights. Thomas Jefferson their arch-enemy was President. He gave no heed to their claims to the disposal of local offices. He had retorted on their clergy for their attacks, that they had " wrapped the Christian religion in rags of their own. . . . Divest it of these, and it is a religion of all others friendly to liberty, science, and the finest expression of the human mind." Political experience there was none for them to draw upon. They acted on the native impulses of their individual characters. They did not wait for a popular reaction as a defeated party would now do ; nor take any steps to deserve such a reaction. Hamilton devised a cunning scheme for a third party, to be called " The Christian Constitutional Party "; hoping to rally the religious element of the nation against free thinking. For at this moment was taking place an alliance of liberty in the church and liberty in the state. Thomas Paine who had been left to his ignominious and undeserved fate by Adams, was sent for by Jefferson and brought home in a national vessel. Every mile that the Puritans moved westward increased their tendency to break with precedent. Ames declared the overthrow of Federalism to be due to the newspapers. They are, he said, " an overmatch for any government." The only possible political devices left to the fallen leaders seemed to be an appeal to religious prejudice ; the censorship of the press ; the renewal of the Sedition and Alien Acts. But all these failed. Federalist judges were impeached by the Re-

publicans ; the tyrannous Acts were suppressed ; religious liberty was encouraged.   " We must be prepared,"
wrote one, " to see the doom of every influential Federalist, and of every man of considerable property who
is not of the reigning sect."

Desperation succeeded discouragement, and desperate
measures followed political scheming.   Judge Reeve of
Connecticut wrote to Tracy in Congress, " I have seen
many of our friends ; and all that I have seen, and most
that I have heard from, believe that we must separate ;
and that this is the most favorable moment."   There
was some effort at secresy ; but the conspirators could
not have hoped not ultimately to be discovered.   They
undoubtedly believed New England was with them ;
and they should be safe.

Pickering was chief conspirator.   Believing, without .
a wavering doubt, in his own political sagacity, he was
unwilling to brook a suggestion of caution or delay.
Cabot was well enough for a philosopher ; but for
action he was too slow and timid for Washington's Secretary of State.   Nor did Pickering have a thought that
Hamilton was as much entitled to leadership as himself.   If delay were tolerated, he insisted that democracy would have its work of ruin accomplished.   The
attitude of Jefferson afforded apparently every advantage for conspirators to, at least, dally with plots.   In
his inaugural he had said, " If there be any among us
who would wish to dissolve the Union, or to change its
republican form, let them stand undisturbed, as monu-
ments of the safety with which error of opinion may be
tolerated where reason is left free to combat it."

Pickering believed the proposition to secede " would
be welcomed in Connecticut, and could we doubt of
New Hampshire ?   But New York must be associated ;

and how is her concurrence to be obtained ? She must be made the centre of the confederacy. Vermont and New Jersey would follow of course, and Rhode Island of necessity." Roger Griswold, examining the finances, had found that the States above mentioned, to be embraced by the "Northern Confederacy," "now pay as much or more of the public revenues, as would discharge their share of the public debts, due those States, and abroad." Ex-Governor Griswold wrote to Oliver Wolcott, "The project which we had formed was to induce if possible the legislatures of the three New England States who remain Federal to commence measures, which should call for a reunion of the northern states." The three States he relied on were Connecticut, Massachusetts then including Maine, and New Hampshire. "The people of the East cannot reconcile their habits, views, and interests," wrote Pickering, "to those of the South and West." George Cabot however wrote that, while "a separation at some period not very remote may probably take place," he thought "a separation now is impracticable. If it is prematurely attempted, those few only will promote it who discern what is hidden from the multitude"; that is the multitude would not feel as the leaders felt who saw power sliding from their grasp. "We shall go," he added, "the way of all governments wholly popular—from bad to worse—until the evils no longer tolerable shall generate their own remedies." Here was clearly treason not only to the union, but to popular government ; and it was evidently the sentiment of a very large class of Federalist leaders. Hamilton at a banquet in New York, expressed his views of popular government by shouting, "The People ! Gentlemen ! the people are a great Beast !" John Adams now

9

watched the traitors with the anxious eagerness of a
detective. It was certain he had discovered more or
less of their purposes ; and was ready at the first move
to pounce on them with genuine Adams' fury. Cabot
wrote from Boston, after consulting Fisher Ames, Chief-
Justice Parsons, and a few more, that, while some were
of the same opinion as Pickering, most thought the
time not quite ready. As for himself he could not believe essential good would come from secession : " while
we retain maxims and principles which all experience
and reason pronounces to be impracticable and absurd.
Even in New England, where there is among the body
of the people more wisdom and virtue than in any other
part of the United States, we are full of errors. We are
too democratic altogether; and I hold democracy to be
the government of the worst. . . . A separation now
is impracticable, because we do not feel the necessity
of it. The separation will be unavoidable when our
loyalty is perceived to be the instrument of impoverishment." In other words Cabot and " the Essex Junto "
saw the country so prosperous under Jefferson that
they dared not precipitate secession. Griswold was in
despair. He wrote to Wolcott that : " whilst we are
waiting for the time to arrive in New England, it is
certain the democracy is making daily inroads on us,
and our means of resistance are lessening every day.
Yet it appears impossible to induce our friends to make
any decisive exertions."

Utterly unable to move ahead without New York,
there was now initiated the most unprincipled plot ever
conceived under a free government ; ending in a fatal-
ity as wretched as the plot itself. New York was rapidly approaching its gubernatorial election. On the
one side was the Federal party, with Morgan Lewis in

nomination—backed by a large number of the Republicans; on the other Burr and his friends. Notwithstanding the fact that Burr had failed of securing the endorsement of Hamilton and the New York Federal caucus, the Federalist leaders of New England, Pickering, Griswold, Wolcott, and others, put their heads together, and agreed to throw all their influence for him, on the understanding that, thus securing New York, he should carry it into the proposed Northern Confederacy. Burr was nominated by a few Republicans Feb. 18, 1804. Griswold wrote: "If Colonel Burr is elevated in New York to the office of governor by the votes of Federalism, will he not be considered, and must he not in fact become the head of the Northern interest? But what else can we do? By supporting Mr. Burr, we gain some support, although it is of a doubtful nature, and of which God knows we have cause enough to be jealous. In short I see nothing else left for us." Pickering with his usual frankness wrote: "The Federalists anxiously desire the election of Mr. Burr. Mr. Burr alone we think can break the Democratic phalanx. And if a separation should be deemed proper, the New England States, New York, and New Jersey would naturally be united." Rufus King was won over substantially in New York. But Hamilton, clearly seeing that such a conspiracy would only end in displacing himself as the great leader of the Federalists, threw all his weight against Burr. it was a sharp battle at the polls; and Burr failed by only seven thousand votes of carrying the State, while he barely carried the city of New York. The plot was killed. The New England conspirators could do nothing whatever but retire into their own States, and leave Burr to the fatal folly of their friendship.

Twice Burr had met Hamilton just at the door of completed ambition. Twice Hamilton had pronounced him a man unworthy of popular confidence. He had, in the recent contest, alluded to him as despicable. Burr certainly had endured bravely ; and for the most part silently. He had fought at great odds. His private character did not compare so unfavorably with that of Hamilton. His views of popular government were not more unsound. Burr was now thoroughly ruined. He had lost his place with the Republicans ; his alliance with the desperate Federal faction had flatly failed. He demanded of Hamilton to retract what he had written and said ; and when the latter refused, he challenged him by the code of honor then current, and shot him. Hamilton became a saint and a martyr ; Burr became as hopeless an outcast as Cain. Wherever he went he went with the brand of a murderer. The Federalists turned on him for killing their ablest leader, whom however they seldom followed ; the Republicans he had taught not to endure him.

Such was the speedy progress and such the end of the great effort to create a Northern Confederacy. Nothing in this matter can be set down to the credit of Hamilton. It was a question whether Burr should supplant him as the leader of the Federals. The large majority of these same Federals however left him to follow Burr. Yet Hamilton did without question believe that the Union was the better way to secure what in common they all aimed after. One of the last letters that he wrote on political affairs says : " Dismemberment of our empire will be a clear sacrifice of great positive advantages without any counterbalancing good ; administering no relief to our real disease, which is democracy—the poison of which, by a subdivision, will only be the more concentrated in each part, and consequently the more virulent." He and Pickering differed only in the means. Hamilton desired as fully as the Essex Junto to weaken popular government; but his imagination looked not to rule over a petty New England confederacy, but over a vast continent. Hamilton, just before the duel said to Colonel Smith, son-in-law of John Adams: " Go to Boston, and see the principal men there. Tell them from me, as my request, for God's sake, to cease these threatenings about the dissolution of the Union. It must be made to hang together as long as it can be made to." Hamilton had the misfortune of not being a Puritan. He could see no charm in provincialism. While working with New Englanders they and he always had opposite aims. What he sought for was a vast empire. What they sought for was non-interference, even in a corner province.

" Hamilton," says Schouler, " was fitted to rule a decaying, not to lead a rising republic." His ambition was undoubtedly to be the Napoleon of a vast western empire. His intellect was imperial; his tastes autocratic. He had however no great power as a leader. He quarreled in turn with every prominent man with whom his lot was cast. As a mere Aid, he had flouted Washington, who petted him and humored him as he humored no other. He violently assailed and abused Adams. He shamelessly maligned Jefferson; and Burr he traduced as a scoundrel. When Lafayette proposed to settle in America Hamilton, while professing pleasure, secretly did all he could to prevent the accession of so influential an ally of the Republicans. It is difficult to speak of the man with all the plainness and severity needful, without seeming to overlook his greatness. He dazzles us with his daring ; and delights us with his precocity. His arrogance even has a touch in it like that of a beautiful but spoiled child. He had created the Federal party, and bottomed its few victories and brief success. He had dictated to Washington, and undertaken to dominate every department of government. When Adams proved an unpliant tool he dismembered the party rather than yield to his chief. More capable of logical words than of logical action ; he was a pessimist at thirty, and threw away his life at forty-five.

Gouverneur Morris his contemporary says: " General Hamilton hated republican government, because he confounded it with democratic government. He believed our administration would be enfeebled progressively at every new election, and become at last contemptible. He never failed on every occasion to advocate the excellence of and avow his attachment to monarchical government." Still his biographers insist that, apart from Jefferson's charges, there is no proof of Hamilton's monarchical desires. One of his most ardent admirers, who quotes with approval Marshall's dictum, " that Hamilton was the greatest man the country has ever seen, always excepting Washington, " does not hesitate to say that " Blinded by passion, Hamilton ruined Adams and the party together." Of his proposal to Governor Jay to circumvent Jefferson's election by arbitrary and partisan change of the electoral law, he says, " A more unscrupulous suggestion it would be difficult to conceive." Of his private life he adds, " Hamilton's passions were his bane, and it was owing to their vehemence, that in moral strength he fell short of his intellectual greatness."

If there is any apology for the duelling code, Hamilton had no excuse for accepting a challenge, whatever Burr may have had for sending it. He stood, when he fought, where his own son, his eldest boy had stood three years before. His blood dyed almost the same spot. It would seem as if a father should have shunned the ill-omened ground ; if he did not abhor the crime that offered a life in the place of an apology. But the politics of the day were desperate with intrigue and dishonor. The duel was a natural appendix. Burr meant to kill. His life had been rendered nugatory, his ambition blighted, his character blackened, his chance for the good as well as the great wilfully destroyed by the man before him. Again and again he had nearly touched the goal of ambition, and Hamilton had tripped him at the post. His party, his people, his State, were no longer his. He was not only despoiled of fame ; he was dishonored, disgraced, and tossed aside unmercifully. But it must not be overlooked that Hamilton's ambitions were as badly wrecked as those of Burr. His scheme for a great war, and a command in which he should outshine Washington, Adams had shattered. His hopes were illusory. He could not achieve the Presidency. He had failed as a power behind the throne. Only the fundholders stood by him, as they have to this day not failed him.

The men morally were not unlike. It has been the fashion to paint Burr in blackest colors. But Hamilton's loves were not spotless. He was compelled to publically confess one of his amours. Both loved their families and were loved by them. In patriotism the burden was not all in favor of Hamilton. Burr was never false to his trust until driven to desperation. His plots, which we shall hereafter still be compelled to rehearse, were, very strangely, only a practical effort

to realize a dream that Hamilton himself had outlined.
So far as the Union was concerned, Hamilton never believed it to be permanent.   Burr was up to 1800, as
patriotic, and as selfish, as his rival.   Circumstances as
well as similarity of character, brought them into personal collision.   Burr saw only Hamilton in his way ;
Hamilton saw in Burr the most immediate difficulty to
prevent his supreme influence in New York.   Above
all he could not tolerate him in the Federal party.
Burr deserved defeat for his alliance with traitors,
whose plots he certainly knew.   But defeat meant
hopeless and utter political obliteration.   It was Burr's
last card as a statesman.   The duel was the only possible conclusion of the struggle.   Hamilton was killed ;
Burr was not benefitted.   The death of the former was
the signal for universal amnesty.   All his faults were
atoned by his miserable taking off.   Not in his case
only has a shot killed a man only to create a saint.
Every New England pulpit rung with eulogy.   The
eloquent Eliphalet Nott furnished the schoolboys of
two generations with arguments for hating Aaron Burr
and apotheosizing Alexander Hamilton.

The most notable feature of this episode in American
history is the fact that the masses remained loyal.   The
Federalists as well as the Republicans were true to the
Union as it was, and to the Constitution.   The popular
vote began to be cast with increasing independence of
the leaders whose ambition had been oligarchical.
There was more than the alliance with Burr to show
that these leaders despised the people.   It was impossible for them to conceal the correspondence that betrayed their contempt for popular government.   The
people were beginning to believe what Jefferson said,
that there had been one revolution in favor of the

rights of man in 1776, and another of like character and equal importance in 1800.

No one can look back to 1803 and gather up the threads of dishonor, and trace the efforts of selfishness to destroy the young nation, while yet in the flush of youth, without deep chagrin and anger. There is no apology for the leaders. There was in all their plotting no great moral purpose ; not even a commercial excuse. Although it was free States against slave States, yet there was no anti-slavery sentiment involved. When the trial hour came two who had been cabinet officers, and two governors, with a chief-justice, and half a dozen Congressmen were plotters to destroy what they were under oath to sustain. The men most trusted were found least ready to sacrifice their own power for the good of the whole. Pickering had proved himself in every way a bigot and a plotter. In Adams' cabinet he had betrayed his superior systematically. Only for one man we should be able to set him down as the meanest man in Amercan history. There was furthermore in these men not even sectional pride. They cared only for personal rule. The intent was to create an oligarchy. They deliberately undertook to pull down what Washington and Jefferson builded, that they might rule in a corner of the ruins.

The spirit of secession was also the spirit of dissolution. A Northern Confederacy would not have held together unless by force of outside pressure. Griswold wrote that such was the jealousy of Massachusetts that it would be necessary she should be allowed not only to lead but to take the initiative. She would be sure not to follow anybody else. '' The magnitude and jealousy of Massachusetts will render it necessary that the operation shall begin there.'' The first action must come

from the legislature of Massachusetts, which would
meet the next summer.   Pickering felt sure this could
be brought about ; but it turned out that in a vote of
fifty-four thousand the Federalists only secured a majority of six thousand.   As the Federalists grew weaker
they were the more discordant ; and while the Pickering party were raging for speedy action, Cabot and
Ames and Higginson and Chief-Justice Parsons were
insistent for delay.   Cabot wrote :   "There is no
energy in the Federal party, and there could be none
manufactured without great hazard of losing the State
government.   Some of our best men in high stations
are kept in office because they forbear to exert any influence, and not because they possess right principles.
They are permitted to have power if they will not use
it."   These "right-principled" rebels, in other words,
were watched very closely ; and although the popular
vote had not quite given up its habit of electing them,
it would swing away very quickly at any overt act of a
traitorous sort.

It must be borne in mind that not once in this plotting of 1803–4 was the right of a State or of a group of
States to secede questioned.   The only arguments
against secession were (1) The immaturity of the plot,
the unripeness of the people for following the leaders,
and (2) The probability that nothing would be gained
by withdrawal into a smaller confederacy.   Hamilton's
only argument against the later phase of the movement
was that the real trouble was not so much Union with
the South, or even the influence of Virginia, but it was
democracy ; and he could not see that by subdivision
of the union any other result would be achieved than to
make in some parts democracy all the more concentrated and troublesome.   Two or three of the leaders

were growing into a conviction that possibly the people were not so very bad repositories of power, and they were trimming sails to catch the popular breeze. Rufus King and Oliver Wolcott were among these. But not one of them argued that neither New England nor New York nor any other section had a *right* to leave the Union. Washington had worked for " an indissoluble Union "; but his colleagues clearly did not suppose the Union indissoluble. Cabot, while unprepared for precipitate action, wished not to be misunderstood. " A separation *now* is impracticable because we do not feel the necessity or utility of it. Separation will be unavoidable when our loyalty to the Union is generally perceived to be the instrument of debasement and impoverishment." This was the average height of the logic used—secession rather than poverty. We look in vain for any high-keyed patriotism. Not a flash anywhere of loyalty to federalism. They will quit the Union if they can, when it costs them too much. The Adams' party was of another spirit ; but they did not argue that secession was impossible ; they proposed to make it hot for those who plotted it. Cabot thought possible that the administration could be involved in a war with Great Britain ; and that such a war would be so unpopular in New England as to give a chance for disunion. Meanwhile they must wait. There really seems to be but one gleam of even tolerable manhood in this whole affair ; the leaders really did not believe in the people's capacity for self-government. They were honest in this. The training in New England had created a vast self-sufficiency of the cultured " set." They questioned sincerely the ability of the people to get on without them. Democracy was not an instinct in America ; it was a lesson learned by years of experience.

APPENDIX TO CHAPTER III

THE TORY REFUGEES IN NEW YORK

One of these refugees, Daniel Cox, wrote : " I have lately brought about a general representation of all the refugees from the respective colonies, which now compose a board, called A Board of Refugees, and of which I have the honor at present to be President. We vote by colonies, and conduct our debates in quite a parliamentary style."

The following document was drawn up by this board or congress to suggest to Parliament a method for reconstructing the colonies : " When America submits to the crown of Great Britain, as will soon happen, if proper measures are not neglected, will not a constitution and government in a manner something like the following be most for the honor, security, peace, and interest of Great Britain, and also for the happiness and safety of America, and most compatible to the spirit and genius of both ; to wit : That the right of taxation of America by the British Parliament be given up. That the several colonies be restored to their former constitutions and forms of government, except in the instances after mentioned. That each colony have a governor and council, appointed by the crown, and a house of representatives to be elected by the freeholders ; inhabitants of the several counties, not more than forty nor less than thirty for a colony, who shall have power to make all necessary laws for the internal government and benefit of each respective colony, that are not repugnant or contradictory to the laws of Great Britain, or the laws of the American parliament, made and enacted to be in force in the colonies for the government, utility, and safety of the whole. That an American parliament be established for all the British colonies on this continent, to consist of a lord-lieutenant, barons, to be created for that purpose, not to exceed at present more than twelve, nor less than eight for each colony, to be appointed by his Majesty out of the freeholders inhabitants of each colony; a House of Commons, not to exceed twelve nor less than eight for each colony, to be elected from the respective houses of representatives for each colony; which parliament, so constituted, to be three branches of legislature of the Northern colony, and to be styled and called the Lord-Lieutenant, the Lords, and Commons of the British colonies in America. That they have the power of enacting laws, in all cases whatsoever, for the general good, benefit, and security of the colonies, and for their mutual safety, both defensive and offensive, against the King's enemies, rebels, etc., proportioning the taxes to be raised in such cases by each colony. The mode for raising the same to be enacted by the general assembly of each colony, which, if refused or neglected, be directed and prescribed by the North American parliament, with power to levy the same. That the laws of the American parliament shall be in force till repealed by his Majesty in council ; and the laws of the several legislatures of the respective colonies be in force till the same be repealed by his Majesty, or made void by an act of the American parliament. That the American parliament have the superintendence and government of the several colleges in North America, most of which have been the grand nurseries of the late rebellion, instilling into the tender minds of youth principles favorable to republican, and against a monarchical government, and other doctrines incompatible to the British constitution.''

The following proclamation was suggested by the same body : " To facilitate a submission instead of a treaty, proceed with the army against the rebels with vigor and spirit, and issue a proclamation containing with a constitution for North America, a pardon to all who lay down their arms, and take the oath of allegiance to his Majesty and his government, excepting as necessary examples of justice ;

First.   The several members of the Continental Congress who have been elected and served as members thereof since the Declaration of Independence.

Second.   All governors, presidents of the supreme executive councils, or of other councils, or of any of the colonies acting under the Congress, or any new and usurped form of government.

Third.   All those who have been by his Majesty appointed of his council in any of the colonies, and since taken an active part in the civil or military department, under the Congress or under any establishment of the rebel government.

Fourth.   All judges who have, since the rebellion, passed sentence of death against any of his Majesty's loyal subjects, for any supposed or real crime, committed or pretended to be committed against any laws enacted by Congress, or by any of the usurped legislatures of the colonies, making the fact criminal for which he was condemned to death.

Fifth.   All commissaries and others who have seized and sold the estates of any of his Majesty's loyal subjects, under any pretence whatever, unless it was done by the consent and orders of the rightful owners ; leaving all such to the mercy of his Majesty, to be granted to those only whose conduct merits mercy ; and hold up the same in the proclamation, if any should issue.

Will it not be proper as well as just to have the
estates of the rebels who are gone out of the King's lines
among the rebels forfeited, confiscated, and sold by
commissioners, to be appointed for that purpose ; and
monies arising on the sales to be applied to the use of
the refugees, to compensate for their sufferings by the
rebels, in case of the parliamentary donation. Will not
the perfidy of France and Spain justify Great Britain
in entering into an alliance with Russia, Prussia, and
other powers, against France and Spain, the common
disturbers of public tranquillity, to take and divide
among them all their islands in the West Indies.''

LETTER WRITTEN BY ALEXANDER HAMILTON TO
GOVERNOR JAY

NEW YORK, May 7, 1800.

DEAR SIR :

You have been informed of the loss of our election in
this city. It is also known that we have been unfortunate throughout Long Island and in Westchester.
According to the returns hitherto, it is too probable
that we lose our senator for this district.

The moral certainty, therefore, is, that there will be
an Anti-federal majority in the ensuing legislature ;
and the very high probability is, that this will bring
Jefferson into the chief magistracy, unless it be prevented by the measure which I shall now submit to
your consideration ; namely, the immediate calling together of the existing legislature.

I am aware that there are weighty objections to the
measure ; but the reasons for it appear to me to out-
weigh the objections ; and, in times like these in which
we live, it will not do to be over scrupulous. It is easy

to sacrifice the substantial interests of society by a strict adherence to ordinary rules.

In observing this I shall not be supposed to mean that any thing ought to be done which integrity will forbid ; but merely that the scruples of delicacy and propriety, as relative to a common course of things, ought to yield to the extraordinary nature of the crisis. They ought not to hinder the taking of a legal and constitutional step to prevent an atheist in religion and a fanatic in politics from getting possession of the helm of state.

You, sir, know in a great degree the Anti-federal party ; but I fear you do not know them as well as I do. 'T is a composition, indeed, of very incongruous materials, but all tending to mischief—some of them to the overthrow of the government, by stripping it of its due energies ; others of them to a revolution after the manner of Bonaparte. I speak from indubitable facts, not from conjectures and inferences. In proportion as the true character of the party is understood, is the force of the considerations which urge to every effort to disappoint it ; and it seems to me that there is a very solemn obligation to employ the means in our power.

The calling of the legislature will have for object the choosing of electors by the people in districts ; this (as Pennsylvania will do nothing) will ensure a majority of votes in the United States for a Federal candidate. The measure will not fail to be approved by all the Federal party, while it will, no doubt, be condemned by the opposite. As to its intrinsic nature, it is justi-fied by unequivocal reasons of public safety.

The reasonable part of the world will, I believe, approve it. They will see it as a proceeding out of the common course, but warranted by the particular nature of the crisis and the great cause of social order.

If done, the motive ought to be frankly avowed. In your communication to the legislature, they ought to be told that temporary circumstances had rendered it probable that, without their interposition, the executive authority of the general government would be transferred to hands hostile to the system heretofore pursued with so much success, and dangerous to the peace, happiness, and order of the country. That under this impression, from facts convincing to your own mind, you had thought it your duty to give the existing legislature an opportunity of deliberating whether it would not be proper to interpose, and endeavor to prevent so great an evil, by referring the choice of electors to the people distributed into districts.

In weighing this suggestion, you will doubtless bear in mind that popular governments must certainly be overturned ; and, while they endure, prove engines of mischief, if one party will call to its aid all the resources which vice can give, and if the other (however pressing the emergency) confines itself within all the ordinary forms of delicacy and decorum.

The legislature can be brought together in three weeks, so that there will be full time for the object ; but none ought to be lost.

Think well, my dear sir, of this proposition ; appreciate the extreme danger of the crisis ; and I am unusually mistaken in my view of the matter if you do not see it right and expedient to adopt the measure.

Respectfully and affectionately yours.

ALEXANDER HAMILTON.

This letter was endorsed by Governor Jay, " Proposing a measure for party purposes which I think it would not become me to adopt."

LETTER WRITTEN BY AARON BURR TO HIS DAUGHTER
THEODOSIA, JUST PREVIOUS TO THE DUEL

NEW YORK, July 10, 1804.

Having lately written my will, and given my private
letters and papers in charge of you, I have no other
direction to give you on the subject but to request you
to burn all such as, if by accident made public, would
injure any person. This is more particularly applicable to the letters of my female correspondents. All my
letters and copies of letters, of which I have retained
copies, are in the six blue boxes. If your husband or
any one else (no one, however, could do it so well as he)
should think it worth while to write a sketch of my life
some materials will be found among these letters.

Tell my dear Natalie that I have not left her anything, for the very good reason that I had nothing to
leave to any one. My estate will just about pay my
debts and no more—I mean, if I should die this year.
If I live a few years, it is probable things may be better.
Give Natalie one of the pictures of me. There are
three in this house ; that of Stewart, and two by Van-
derlyn. Give her any other little tokens she may
desire. One of those pictures, also, I pray you to give
to Doctor Eustis. To Bartow something—what you
please.

I pray you and your husband to convey to Peggy
the small lot, not numbered, which is the fourth article
mentioned in my list of property. It is worth about
two hundred and fifty dollars. Give her also fifty
dollars in cash as a reward for her fidelity. Dispose of
Nancy as you please. She is honest, robust, and good-
tempered. Peter is the most intelligent and best-dis-
posed black I have ever known. (I mean the black

boy I bought last fall from Mr. Turnbull.) I advise you, by all means, to keep him as the valet of your son. Persuade Peggy to live with you, if you can.

I have desired that my wearing apparel be given to Frederic. Give him also a sword or pair of pistols.

Burn immediately a small bundle, tied with a red string, which you will find in the little flat writing-case —that which we used with the curricle. The bundle is marked " Put."

The letters of Clara (the greater part of them) are tied up in a white handkerchief, which you will find in the blue box No. 5. You may hand them to Mari, if you please. My letters to Clara are in the same bundle. You, and by-and-by Aaron Burr Alston, may laugh at gamp when you look over this nonsense.

Many of the letters of Clara will be found among my ordinary letters, filed and marked, sometimes " Clara," sometimes " L."

I am indebted to you, my dearest Theodosia, for a very great portion of the happiness which I have enjoyed in this life. You have completely satisfied all that my heart and affections had hoped or even wished. With a little more perseverence, determination, and industry, you will obtain all that my ambition or vanity had fondly imagined. Let your son have occasion to be proud that he had a mother. Adieu. Adieu.

A. BURR.

LETTER OF ALEXANDER HAMILTON WRITTEN TO NA-
THANIEL PENDLETON ON THE DAY BEFORE THE
DUEL

On my expected interview with Colonel Burr, I think it proper to make some remarks explanatory of my conduct, motives, and views.

I was certainly desirous of avoiding this interview for the most cogent reasons.

1. My religious and moral principles are strongly opposed to the practice of duelling, and it would ever give me pain to be obliged to shed the blood of a fellow-creature in a private combat forbidden by the laws.

2. My wife and children are extremely dear to me, and my life is of the utmost importance to them in various views.

3. I feel a sense of obligation toward my creditors ; who, in case of accident to me, by the forced sale of my property, may be in some degree sufferers. I did not think myself at liberty, as a man of probity, lightly to expose them to this hazard.

4. I am conscious of no ill-will to Colonel Burr, distinct from political opposition, which, as I trust, has proceeded from pure and upright motives.

Lastly, I shall hazard much, and can possibly gain nothing by the issue of the interview.

But it was, as I conceive, impossible for me to avoid it. There were intrinsic difficulties in the thing, and artificial embarrassments from the manner of proceeding on the part of Colonel Burr.

Intrinsic, because it is not to be denied that my animadversions on the political principles, character, and views of Colonel Burr have been extremely severe ; and, on different occasions, I, in common with many others, have made very unfavorable criticisms on particular instances of the private conduct of this gentleman.

In proportion as these impressions were entertained with sincerity, and uttered with motives and for purposes which might appear to me commendable, would be the difficulty (until they could be removed by evidence of their being erroneous) of explanation or apology. The disavowal required of me by Colonel Burr, in a general and definite form, was out of my power, if it had really been proper for me to submit to be so questioned ; but I was sincerely of the opinion that this could not be ; and in this opinion I was confirmed by that of a very moderate and judicious friend whom I consulted. Besides that, Colonel Burr appeared to me to assume, in the first instance, a tone unnecessarily peremptory and menacing, and, in the second, positively offensive. Yet I wished, as far as might be practicable, to leave a door open for accommodation. This, I think, will be inferred from the written communications made by me and by my direction, and would be confirmed by the conversations between Mr. Van Ness and myself which arose out of the subject.

I am not sure, whether, under all the circumstances, I did not go further in the attempt to accommodate than a punctilious delicacy will justify. If so, I hope the motives I have stated will excuse me.

It is not my design, by what I have said, to affix any odium on the character of Colonel Burr in this case. He doubtless has heard of animadversions of mine which bore very hard upon him ; and it is probable that, as usual, they were accompanied with some falsehoods. He may have supposed himself under a necessity of acting as he has done. I hope the grounds of his proceeding have been such as ought to satisfy his own conscience.

I trust, at the same time, that the world will do me the justice to believe that I have not censured him on light grounds nor from unworthy inducements. I certainly have had strong reasons for what I have said, though it is possible that in some particulars I have been influenced by misconstruction or misinformation. It is also my ardent wish that I may have been more mistaken than I think I have been, and that he, by his future conduct, may show himself worthy of all confidence and esteem, and prove an ornament and blessing to the country.

As well, because it is possible that I may have injured Colonel Burr, however convinced myself that my opinions and declarations have been well founded, as from my general principles and temper in relation to similar affairs, I have resolved, if our interview is conducted in the usual manner, and it pleases God to give me the opportunity, to reserve and throw away my first fire, and I have thoughts even of reserving my second fire, and thus giving a double opportunity to Colonel Burr to pause and to reflect.

It is not, however, my intention to enter into any explanations on the ground—apology, from principle, I hope, rather than pride, is out of the question.

To those who, with me, abhorring the practice of duelling, may think that I ought on no account to add to the number of bad examples, I answer, that my relative situation, as well in public as private, enforcing all the considerations which men of the world denominate honor, imposed on me (as I thought) a peculiar necessity not to decline the call. The ability to be in future useful, whether in resisting mischief, or effecting good, in these crises of our public affairs which seem likely to happen, would probably be inseparable from a conformity with prejudice in this particular.   A. H.

HARRIET MARTINEAU ON DUELLING IN AMERICA

" The manners of the Americans (in America) are the best I ever saw ; and these are seen to the greatest

advantage in their homes ; and as to the gentlemen in
travelling. The hospitality of the country is cele-
brated ; but I speak now of more than usually meets
the eye of a stranger ; of the family manners which
travellers have rarely leisure or opportunity to observe.
It is not so much the outward plenty, or the mutual
freedom, or the simplicity of manners, or the incessant
play of humor, which characterizes the whole people,
as the sweet temper which is diffused like sunshine over
the land. I imagine that the practice of forbearance
requisite in a republic is answerable for this pleasant
peculiarity. In a republic no man can in theory over-
bear his neighbor ; nor as he values his own rights can
he do it much or long in practice.

Some may find it difficult to reconcile this prevalence
of good temper with the amount of duelling in the
United States ; with the recklessness of life, which is
not confined to the semi-barbarous parts of the country.
In New Orleans there were fought in 1834 more duels
than there are days in the year ; fifteen on one Sunday
morning ; in 1835 there were 102 duels fought in that
city between the first of January and the end of April.
When the world numbers the duel between Clay and
Randolph ; that Hamilton fell in a duel ; and several
more such instances ; there may be wonder that a nation where such things happen should be remarkably
good tempered. But New Orleans is no rule for any
place but itself. But even on that spot there is some
feeling of disgust and shame at the extent of the practice. A Court of Honor was instituted for the restraint
of the practice. Its function degenerated into choosing
weapons for the combatants. Hamilton's fate opened
men's eyes to the philosophy of duelling, and is working to that purpose more and more. At the time it was

pretty generally agreed he could not help fighting, now there are few who think so. His correspondence with his murderer previous to the duel is remarkable. Having been told that Hamilton was the 'greatest man' of the country, I was interested to see what a greater than Washington could say in excuse for risking his life in so paltry a way. I read his correspondence with Colonel Burr with pain. There is fear in every line of it— a complicated, disgraceful fear. He was obviously perishing between two fears; of losing his life, and of not being able to guard his own honor against the attacks of a ruffian. Between these two fears he fell. I was talking over the correspondence with a duelling gentleman. ' O,' said he, ' Hamilton went out like a Capuchin.' So the greatest man did not obtain even that for which he threw away what he knew was considered the most valuable life in the country. This is as it should be. When contempt becomes the wages of slavery to a false idea of honor, it will cease to stand in the way of the true ; and ' greatest men ' will not end their lives in littleness."—*Travels in the West.*

CHAPTER IV

THE revolution from Federalism soon became complete. In 1800 New England voted as a unit against Jefferson; while the leaders denounced him as an atheist and a political libertine. Ruin was prophesied; both in state and church. But in 1804 he carried Massachusetts, New Hampshire, Vermont, and Rhode Island; while barely losing Connecticut. In fact every State in the Union except Delaware and Connecticut, with two elevenths of Maryland, had voted solidly for the man who stood for reform, economy, simplicity; and who above all trusted the people. The people trusted him. Their love and confidence grew into fanaticism. In a quiet undictatorial manner he was a real dictator. He had written the Declaration of Independence, and it had been said he was a theorist; but he had proved pre-eminently practical. The taxes went down, but the debt went down too. He with Mason, Henry, and Madison had destroyed the State church and primogeniture in Virginia; and now he had broken the power of the Church State which Federalism had created. Slowly the territorial instability of the Union was being reversed. The effort of New

153

England leaders to break off the North failed. The different sections were drawing closer together. Roads were being constructed that served as bonds. New England had pioneered its way into western New York, and around Lake Erie into what are now Ohio, Michigan, Indiana, and Illinois. The tide westward was even more strong at the South. These new settlements loved the old as children love parents. A few more years and every section would so enlarge its ties as to lose its sectionalism.

One only chance now remained for a breach in the fellowship of States. We have seen how in 1803 Jefferson, negotiating with France for so much of Louisiana as would give us free navigation of the Mississippi, had had flung at him the astounding offer of the whole territory of Louisiana. This was an unknown quantity, but it was certainly a vast area. No one knew its exact size—least of all France.

So long as any foreign power held Louisiana it was clear that we could not utilize our western territories. We must own the Mississippi as the natural commercial outlet of all the productions of our West and South. The Ohio would flow in vain if we must stop free navigation at its mouth. While New Orleans was in the hands of Spain Hamilton insisted that it should be seized by force of arms. Jefferson was averse to war, and sent Monroe on a special mission to France, to whom Spain had transferred Louisiana. To him he said, " On the event of this mission depends the future destinies of this Republic." Napoleon had secured Louisiana from Spain with a definite plan of sending a force to this country to recover Cuba, and all our West that had originally been acquired by the French. What would have been the consequences to our States

we can only conjecture.   But England at this juncture
threatened to renew war with France.   Napoleon had
his hands full at home.   He also needed money.
Turning to our commissioners, Livingston and Monroe, he proposed to sell the whole of his new acquisition to the United States.   It would be a good stroke.
He would make us his friends ; he would replenish his
treasury ; and he could concentrate his forces and his
purposes on European affairs.   Seeking only to secure
the Mississippi as our western boundary, we were now
told to accustom ourselves to the ownership of a vast
area reaching to the Rocky Mountains.   Our new mission must be one of vast migration and settlement.   It
would take all our genius for a century to invent rapid
methods of intercommunication and transmission of
products.   The steam age was fortunately just ahead.

While closing the bargain with Napoleon, Jefferson
saw a possible danger ahead.   The Mississippi might
hereafter become a dividing line between two great
republics or confederacies.   However, he said, we must
go ahead.   ''The future inhabitants of the Atlantic
and Mississippi States will be our children.   We leave
them in distinct, but bordering establishments.   We
think we see their happiness in their union.''

We have seen the consequences of Burr's clash with
Hamilton.   He had crossed the track of the most
petted man in America ; a man whose ambition had
heretofore only been thwarted by Adams ; whom in
return he execrated without fear of consequences.   On
Burr he had taken speedy revenge by preventing him
from supplanting Jefferson.   The fatal step of Burr was
allowing his political enemies to use his name in such
a dishonorable contest.   Event followed event with
fatal readiness, until Hamilton lay in his grave, and

Burr was an outlaw. The fate of selfish ambition had never a more brilliant illustration. Burr's allies who had led him into his pit were indecently unjust. They had not a word of pity for him and his ruin. They cared only for their own fallen leader Hamilton. Those who had been willing, for their piques and ambitions, to make Burr chief of a seceded confederacy, deserted him in a body. Republicans justly hated him ; Federals left him to his fate.

He did not dare to appear in New York, nor indeed long to remain secluded at his own home. As he says in one of his letters, " Two States were rivals for the honor of hanging Aaron Burr." He went as speedily and secretly as possible to Philadelphia ; and there he held a conference with the English minister.

It is certain the daring exploit was already shaped in his mind to break up the Union and rule on the ruins. Mr. Merry, the English minister, wrote at once to the British cabinet, " I have just received an offer from Mr. Burr, the actual Vice-President of the United States —which office he is soon to resign—to lend his assistance to his Majesty's government in any manner in which they may think fit to employ him, particularly in endeavoring to effect a separation of the western part of the United States from that which lies between the Atlantic and the mountains. If after what is generally known of the profligacy of Mr. Burr's character, his Majesty's ministers should think proper to listen to his offer, his present position in this country, where he still preserves connections with some people of influence, added to his great ambition and spirit of revenge, may possibly induce him to exert the talents and ability which he possesses with fidelity to his employers." Burr was fond of ciphers in correspondence, and

of go-betweens in negotiations.   At this time he employed a man named Williamson.   This was in August,
1804.   The next spring Burr would cease to be Vice-
President in March, 1805.   Meanwhile he had gone as
fast as possible to the South where he was not only
safe from arrest, but where he could sow the seeds of
treasonable discontent.   In one of his letters, which
are very cautious, he lets slip that the B—— were
"openly his partizans."   From the South he hastened
back to Washington, where he conferred with several
western men, and took into his confidence Gen. Dayton, Senator from New Jersey.   But everything was so
far merely the purpose of a desperate man to do mischief and win personal glory.   Wilkinson, our senior
general in command of the Mississippi Valley, with
3500 men, set him on a more definite line of work.
According to Wilkinson the new Louisiana purchase,
and especially New Orleans, was the seat of rebellious
discontent; Burr had only to appear there at the head
of a small force, to make himself master of the city and
the valley above.   Wilkinson would go over to him
with the army.   But as preliminaries, Tennessee, Kentucky, and Ohio were to be visited and their friendship won.   A great confederacy should be formed ;
with New Orleans as capital, and the Ohio and Mississippi Valleys as contributory.

Burr was first of all to make a visiting tour of the
States included in the plot, and use his magnetic power
in winning adherents.   Before starting he went again
to the British minister.   Mr. Merry wrote home, "Burr
must have money and a navy."

That the purposes of Burr cannot be made out with
exactness at this juncture, is probably owing to the
fact that he had no one definite plan ; or rather that he

had a dozen. It is certain that he talked to each one
whom he hoped to win, in a strain likely to please the
listener. Setting out for the Southwest he met General Dayton at Pittsburg, with whom he conferred.
He visited Jackson at Nashville, and had an enthusiastic reception. Burr recounts this with great glee in a
letter to his daughter. Jackson afterwards received a
second visit ; but wrote to Burr, at a later date, that
there were rumors afloat that his purposes involved
hostility to the United States ; if that were true he
would hold no conference on the subject ; but if untrue,
and his intentions were to proceed against Mexico and
Spanish America, he would "join him with his whole
division of the army." The latter, as well as some
others, were probably let into the secret no farther than
to believe that Burr proposed to drive the Spaniards
out of Louisiana and Florida. Whether he intended a
second American republic in the Southwest, or a monarchy, did not much trouble these haters of Spanish
rule. The intention was however to dismember the
Union. " Kentucky, Tennessee, the State of Ohio,
with part of Georgia and Carolina, are to be bribed with
the plunder of the Spanish countries west of us, to
separate from the Union." So wrote Clark of New
Orleans, who was in Burr's confidence. The French
minister wrote to his government, " The project of
effecting a separation between the Western and Atlantic States marches on. · The division of the Confederate
States seems to be inevitable—assuming it to take place,
should we not have a better chance to withdraw, if not
both confederations, at least one of them from the yoke
of England ? " England rejoiced in the proposed destruction of the Union ; France saw no harm to itself
in such an event. Turreau wrote further to Talley-

rand, '' Mr. Burr's career is generally looked upon as
finished ; but he is far from sharing that opinion, I believe he would rather sacrifice the interests of his
country than renounce celebrity and fortune. Louisiana is to become the theatre of his intrigues ; he is
going there under the ægis of General Wilkinson.''

It is curious that France and the United States
should have produced so many parallels ; most curious
perhaps was it that Burr and Talleyrand should have
been brought into contact—the two political Iagos of
the 19th century. Talleyrand combined all the vices
possible to a man of great intellect. He was mercenary
to a degree surpassed only by his lechery. He was
treacherous to every party and every regime. It is impossible with all this not to allow that he was a most
remarkable statesman, with a keen insight into affairs
and a political judgment unsurpassed in France. He
was probably in his unconscious self a good deal of a
democrat ; but by conscious policy he was whatever
was winning. Burr was as near to being his counterpart as could be produced by peoples differing as widely
as the Saxon and Frank. They will live as buffets
against which all that is honorable, loyal, patriotic,
and unselfish shall appear more admirable in a public
man. It was Burr's fate however to be the petitioner.
It was Talleyrand's fate to spurn him.

Here was on the whole a pretty kettle of fish. The
creoles of Louisiana were restive under their transfer
from Spanish and French to American authority.
They were to be tickled with the idea of independence,
and a league direct with Great Britain. England, who
had by no means come to consider us a full-fledged
nation, did not hold it to be contemptible to allow her
minister to plot with conspirators. Pitt and Lord Mel-

ville were in constant conference with Burr's London friend Colonel Williamson. Dayton of New Jersey, just out of the Senate, shared in all the counsels ; while other men in high authority, including at least two Senators, and probably three, were ready to aid the scheme. Wilkinson was to be second always to Burr. New Orleans was to be the capital. The boundary proposed at first was, on the east, the Alleghany Mountains. But on the west they would conquer and annex Mexico. So the Burr Empire would be the very heart and soul of the continent; leaving only provinces outlying. Of England, Burr asked " two or three ships of the line, the same number of frigates ; and a proportional number of small vessels." Besides he asked of Pitt a loan of one hundred and ten thousand pounds, to be given in the name of two of Burr's richest friends, John Barclay of Philadelphia and Daniel Clark of New Orleans.

But Burr did not count alone on the success of his scheme in creating a Southwestern confederacy. He had seen too much of the Federalist leaders to believe that the end of conspiracy had been reached in New England. The English minister wrote : " Burr observed (what I readily conceive may happen) that when once Louisiana and the western country become independent, the Eastern States will separate themselves from the Southern ; and thus the immense power which has now risen up with so much rapidity in the western hemisphere, will by such a division, be rendered at once informidable." But Burr mistook the drift in New England, and the popular sentiment, which was setting strongly the other way. His own reckless act in shooting Hamilton had alienated those who had conspired with him ; and now for Burr and whatever he might propose they had only horror. Should he try to dismember the Union, the Union would grow only the stronger in New England and New York.

Burr's project reached out without doubt to a scheme for securing the alliance of the Northwest. Wilkinson gave him a letter to Governor Harrison of the Indian Territory ; asking Harrison to secure the return of Burr to Congress as Territorial delegate—a boon " on which the Union may much depend." The object seems to have been to give Burr influence with the Northwest, and a Congressional lever. The secession thus would have covered the whole Mississippi Valley from New Orleans to the Lakes. But nothing came of his visit to Harrison. The Northwest was filled with New Englanders ; and the hatred of Burr was as strong in western New York, the Western Reserve, the Marietta and Michigan settlements, as any sentiment that could penetrate the souls of an honest people. There never has been a tinge of disloyalty in the history of these pioneers, that, leaving New England for the wilderness, left behind the selfish " Best."

Burr and his confidants were meanwhile flitting about the West, and using the press as far as they dared. Blennerhassett, an enthusiastic, poetic Celt, whom he found in a charming retreat in the Ohio, where he had turned an island into a paradise, became a devotee. Wilkinson was a thoroughly weak creature and a coward. He had sounded his officers, and had found them true as steel to the Union. There is not on record that there was one traitor in the army except Wilkinson himself. At their next meeting Burr found Wilkinson with no stomach for the enterprise. Burr urged that the people were ripe for revolt. Wilkinson answered, " My friend, if you have not profited more

11

by your journey in other respects than in this you would better have stayed at Washington. Surely no person was ever more mistaken. The Western people disaffected to the Government ! They are bigoted to Jefferson and Democracy."

Burr hastened to Washington. No hope from England ! none from France ! Only a desperate man would have tried his next venture. He sent Dayton to sound the Spanish minister. The purpose was to alarm the Spaniards into paying for secrets concerning " plots engaged in by England and Burr, against Spanish dependencies." Yrugo listened and secured the whole plot ; but saw through its meshes at a glance.

But as Burr's scheme unfolded, and the web wove, he was not far from a scheme even more infamous and insurrectionary. Finding his hope of securing financial aid, and especially ships from the British government disappointed, he planned, as his intimate Dayton explained in full, the capture of Washington and the overthrow of the government. He proposed to introduce into Washington a number of men, well armed, but apparently honest folk, who at a signal should seize the President and Vice-President and president of the Senate. They should then take the arsenal ; and Burr would appear, to establish a new military government. But in case of serious hindrance he would burn the national vessels, except what he wished for his own use. On these he would embark with his followers for New Orleans, and there declare a new Confederacy and establish a government. There was evident at this stage the despair of unmasked scoundrels.

There is no question but the government was fairly well posted in Burr's movements and schemes. It was

waiting for overt acts.   Jefferson however felt sure of
the people ; or professed to do so.   It is probable he
also knew the full danger of handling the conspirators.
He had a marvellous power of letting mischief work out
its own ruin.   Burr had turned to England, to France,
to Spain ; and now was growing desperate.   He was
out of funds.   Dayton also, although squeezing a few
thousands out of Spain for " information," was penni-
less.   The extreme of desperation appears in this communication to Eaton, " If he could gain over the
marine corps, and secure to his interests the naval commanders, Truxton, Preble, Decatur, he would turn
Congress out of doors ; assassinate the President ; and
declare himself Protector of an energetic government."
The idea of winning Decatur and Preble to a crazy act
of rebellion was the dream of a man no longer having
a foot of ground to stand on.

Diplomacy failing in every direction, and his case
growing steadily weaker, Burr engaged in a game of
bluff.   Writing a dispatch full of lies to Wilkinson, he
set out in August of 1806 for New Orleans.   Landing
at Blennerhassett's Island, " the enthusiastic Celt was
assured that before the end of the year Louisiana would
be independent, with Burr for its ruler, and under the
protection of England.   Wilkinson and the United
States army were pledged to accept the revolution, and
support Burr."   Burr then crowded on to Nashville,
where he was tendered a public dinner.   He had
scarcely left Nashville, when a proclamation signed
" Andrew Jackson, Major-General Second Division,"
dated Oct. 4, 1806, was issued, ordering all brigade
commanders to place their brigades on a footing to move
at shortest notice.   The Spanish minister wrote to his
government a very minute detail of Burr's movements

and projects. " Recruits were gathering at Blenner-hassett's Island." " In passing Cincinnati they expect to get possession of five thousand stand of arms deposited there by the government." " I understand Colonel Burr has already written a declaration of independence (for the seceding States) couched in the same terms that the States adopted against Great Britain."

Burr began building boats on the Ohio, and enlisting men. November 3d he sent three thousand dollars to Jackson, with orders for five large boats, supplies and recruits. But the people were rapidly waking up to what was going on. Foolish talk had been tolerated ; but here was overt action. Burr was soon warned, and alarmed. He began to declare his innocence. Jefferson believed the time had come for governmental action. John Graham was sent as a trusted agent on Burr's trail. The Ohio legislature took action for calling out the militia. Kentucky courts summoned Burr to answer charges for violating United States' laws, by setting on foot a military expedition. He was however acquitted. Wilkinson became thoroughly alarmed, and played double. Jackson and other friends became suspicious that Burr had never disclosed to them the whole of his designs. History however will never be able to make this episode in Jackson's life satisfactory to honest and loyal citizens. The governor of Ohio took measures to seize the boats building by Burr on the Ohio. But the arch-conspirator succeeded in entering the Mississippi. There, landing at a small town, he came on a newspaper slip showing that General Wilkinson had deserted him, and to save himself from court-martial, was playing the role of savior of his country. He was indeed already denouncing Burr as a traitor, and writing voluble letters to Washington.

Burr stopped his flotilla ; sank his cases of guns in the river, and surrendered to the governor of the Mississippi Territory. The governor acted with promptitude. The jury that was summoned declined to enter an indictment. Burr took his chances and fled into the woods. After about a month he was discovered in a cabin and arrested.

In March 1807 he was forwarded to Richmond for trial on a charge of " (1) High misdemeanor in setting on foot within the United States a military expedition against the dominions of the king of Spain, a foreign prince with whom the United States at the time of the offence were, and still are, at peace. (2) Treason in assembling an armed force with a design to seize the City of New Orleans, to revolutionize the territory attached to it, and to separate the Western from the Eastern States." The defense urged that there was " no evidence of treason—that there was nothing like an overt act, or probable ground to believe him guilty —that the letter in cypher to General Wilkinson was not delivered by Mr. Burr. As to the other point the evidence was trivial "—" that a criminal act must be proved—that as treason was of all crimes most heinous, it required the strongest evidence to support it—that his flight was not from justice but from military oppression." It was pointed out that there were no soldiers anywhere to prove an expedition. Burr took a prominent part in his own defense ; and it was with his usual shrewdness and plausibility.

Chief-Justice Marshall was a confirmed enemy of Jefferson. He had openly expressed his belief that the latter was " unqualified for the Presidency." Jefferson's administration had attacked the Federal judiciary, of which Marshall was chief. Two judges had been impeached.   Never in our history did the two departments so sharply collide as just at this period.   Marshall with all his learning and poise was thoroughly human.   His rulings were such that it was impossible to convict Burr under the charges as formulated.   Jefferson had allowed Burr to go as far as the public safety would allow ; but even as it was, Marshall would not allow that any overt act of treason or even misdemeanor had been proven.   The trial was made to turn into a personal attack upon Mr. Jefferson.   Every opportunity was given to dishonor him.   He was cited to appear personally on the witness stand, to be badgered by Burr and his counsel.   He refused to appear.   Although the evidence now known was but a moiety of it then obtainable, it was evident to every one that Burr was guilty.   Marshall's rulings however led the jury to believe they could not convict.   The indictment went back, after a few hours endorsed " We of the jury say that Aaron Burr is not proved to be guilty under *this* indictment by any evidence submitted to us; we therefore find him not guilty."

If the Judiciary, in John Marshall, at this time was asserting its position, and gradually assuming its place as equal in importance to the other departments of government, it was doing much more.   It was undertaking to degrade the Executive.   If the President could be compelled by subpœna to attend any and every court in the United States, where the lawyers chose to require his presence as witness, we should be in a curious predicament.   But John Marshall decided the President could be cited, and that he must obey.   Jefferson flatly refused to allow the judiciary to enter on another course of dictatorial usurpation.   The trial of Judge Chase was too recent.

The verdict was probably just as well for the country. It got rid of Burr. It was not necessary to have a political hanging. It left no chance to make a martyr of the traitor. Jefferson was however unreconciled to the result. "The framers of our Constitution certainly supposed they had guarded as well their government against destruction by treason as their citizens against oppression under pretense of it."

The trial showed Burr to be of the same indomitable will he had always manifested, and quite as characterless. His amours were carried on under the nose of the court. He had so committed many of the witnesses in his plots, that they dared not swear the whole truth. He openly avowed his intention, if released, of immediately starting for England to collect money to pursue his plans in America. These plans he persistently insisted were wholly innocent.

Burr was not at all pleased with the decision of Marshall, or with the verdict of the jury. His note to his daughter Theodosia reads : " After all, this is a drawn battle. The Chief-Justice gave his opinion on Tuesday. After declaring that there were no grounds of suspicion as to the treason, he directed that Burr and Blennerhassett should give bail in 3000 dollars for further trial in Ohio. This opinion was a matter of regret and surprise to the friends of the Chief-Justice, and of ridicule to his enemies." Mr. Hay (counsel for prosecution) immediately said that he should advise the government to desist from further prosecution. There was however a supplementary trial in the second court; and after that Burr was bound over to answer to the State of Ohio. He never appeared, and no steps were taken to secure him.

During this episode in American history the one

most infinitesimally contemptible character did not prove to be Burr, but Wilkinson. Senior general of the United States, he was a pensioner of Spain for twenty years—receiving his annual stipend from their most Christian majesty ; and acting as spy and traitor throughout. That he should also have been an accomplice of Burr was a bagatelle. That at last he betrayed Burr was quite a necessity for such a nature; for could he be true to anything or anybody ? It is not at all unlikely that he had acted for Spain all along ; drawing Burr on to break up the Union ; but when Burr's schemes reached out to attack Mexico, his Spanish masters were convinced the States would be less dangerous to them than Burr would be. Then their contemptible tool denounced his ally to our government ; making good his own safety by playing the natural part of informer and witness. The Spanish minister wrote : '' Wilkinson detests this government, and the separation of the Western States has been his favorite plan. Burr persisted in a wild project against Mexico. Wilkinson foresaw he would lose the generous pension and honorable employment he enjoys from our king. In such an alternative he has acted as was to be expected, that is he has sacrificd Burr in order to obtain on the ruins of Burr's reputation the advantages I have pointed out.'' The scoundrel posed as savior of the Union (as he really was); held on to his two thousand a year from the Spanish king ; secured immunity from his own double treason ; and stood on the witness block to try to get Burr hanged. John Randolph wrote : '' Wilkinson is the only man that I ever saw who was from the back to the very core a villain. Perhaps you never saw human nature in so degraded a situation as in the presence of Wilkinson before the grand jury.''

General Jackson went to Richmond; and while having a taste of sympathy for Burr, harangued himself into a fury about Wilkinson. It was evident he would have enjoyed hanging him with his own hands. Wilkinson at a later date sent an agent to the Mexican government, commissioned to make known the great pecuniary losses he had encountered in preventing an invasion of that State by the Vice-President of the United States, and to solicit reimbursement to the amount of two hundred thousand dollars. " I," wrote Wilkinson, " like Leonidas boldly threw myself in the pass." But the viceroy declined " to give a single ducat " ; and took immediate steps to have the fellow got out of the country. Wilkinson was evidently not satisfied with the annual stipend he received from Spain. Although court-martialed he was not cashiered. When in 1812 he was assigned to command on the northern lakes his incompetence and cowardice led to another court-martial. General Scott brushed him aside as " that unprincipled imbecile." He spent his old age on an estate in Mexico bought with the profits of his treason. There he wrote a much more reputable *Life* than he had succeeded in living.

Other characters flickered about the expiring candle of the maddened statesman ; and were singed into oblivion. Among them were the three United States Senators, Dayton of New Jersey, a relative ; Smith of Ohio ; and Adair of Kentucky, the predecessor of Henry Clay.

Burr hastened to secrete himself after the trial ; and as soon as possible took ship for England. Here he certainly made every conceivable effort to approach the government. But England had at that time lost her taste for American adventurers. Canning positively declined to consider any proposition from him. Reports to the contrary however reached America. In October 1808 Jefferson wrote : " Burr is in London, and is giving out that the government offers him two millions of dollars the moment he can raise an ensign of rebellion as big as a handkerchief. Some of his partisans believe this, because they wish it. But those who know him best will not believe it the more because he says it." But in private he had not lost his marvellous magnetism. He succeeded in forming the acquaintance of Jeremy Bentham. In a letter he boasts, " (N. B. Three servants at my command.) In his library I am now writing." Soon the English government obtained enough information of his purposes and talk to order him back to London. There his papers were seized, and he was shut up in prison for a couple of days. Lord Liverpool then bundled him impromptu out of the country. Reaching France, Napoleon ordered him closely watched. He was now in a trap. Striving to get at the emperor he was everywhere rebuffed, treated with Arctic coolness, and not even allowed to leave France. He sent pleading memorials to Napoleon; and they elicited no response. He then turned in every direction to get a passport to allow him to go back to America. In his diary of November 1811 he wrote, " Nothing from America, and really I shall starve. Borrowed three francs to-day." " Dec. 1, 1810. . . . Came on me just as I was out of bed for 27 livres, which took literally my last sous." Finally in July 1811 Napoleon consented to let him leave for America. But the ship was captured by the British, and he was carried to England. Nobody paid any attention to him. He was half starved, but could not get away till March of 1812. He then writes, " I

shake the dust off my feet.   Adieu John Bull.   *Insula
inhospitabilis;* as you were thus called 1800 years
ago.''

Back in New York at last, Burr's ambitions were for
a moment confined to getting food.   He was too poor
to buy a book, or open an office without aid.   This
aid was meagre.   But he had at last an office.   At this
very juncture word came of the death of his grandson,
and that his heart-broken daughter had taken ship
from South Carolina to go to him in New York.   The
ship never reached New York.   It was thought that
a pirate captured her, and that Theodosia was mur-
dered—it is possible the ship foundered in a gale.

There was one lingering political hope in Burr's
mind.   He hoped that Jackson would be nominated for
President in 1815 instead of Monroe.   He wrote malicious letters urging his son-in-law Allston, then governor of South Carolina, to prove himself a man by
opposing the nomination of Monroe.

Still plotting, he secured a commission from Venezuela to raise troops to aid that republic '' for the cause
of liberty.''   But the commission came with the strict
injunction that all who enlisted must look for no more
compensation than bare subsistence.   Under such conditions Burr did not raise his troop.   He had a few
pettifogging cases ; enough to save him from destitu-
tion.   At seventy-eight, worn out with sensuality and
the results of disordered ambition, he married a rich
widow ; but she soon deserted him.   A stroke of
paralysis made him utterly helpless.   He could not
dress himself, nor take a step without assistance.
Without a brother, sister, wife, or child he dragged out
a few years in bitter penury and misery ; and then died
to be buried at public expense.   The Faculty of Prince-

ton College took charge of his body, in kindly recollec-
tion of his father and his grandfather, who had been
Presidents of that institution.

It is impossible while we retrace this damnable record
of a traitor, not to have a sympathetic longing that at
some point in his life the switch might have been turned,
that would have run his engine upon a track of honor
and usefulness, if not have transformed him into a
national ideal. Washington, while honoring Hamilton, was always prejudiced against Burr. Adams proposed to Washington that he make Burr a Brigadier-
General. He reports Washington as replying, "By
all I have seen and heard Colonel Burr is a brave and
able officer ; but the question is whether he has not
equal talents at intrigue." Madison and Monroe at
a later date urged him to appoint Burr minister to
France ; but Washington steadily refused, saying " I
will appoint you Mr. Madison, or you Mr. Monroe,
but I have no confidence in Mr. Burr's integrity."
Was this the keen insight of a great mind ; or was it
the result of the enmity of Hamilton ? At all events
Hamilton proved to be the constant barrier to Burr's
honorable ambition ; with his death nothing remained
for ambition but dishonor.

Was the conspiracy of Burr ever really formidable ?
Of this there can be little doubt with those who have
any knowledge of the peculiar character of the settlers
of the Southwest. The agents of Spain had invited
Kentucky twenty years previously to sever connection
with the States, and enjoy free navigation of the Mississippi in allegiance to Spain. Don Gardoqui was
authorized to treat separately with any of the Western
States. Great Britain had not only done much to make
New England restless, but had never hesitated to

foment disturbance in the South and West.   The later
history of Texas throws a light backward upon Louisiana.   The fact that the major-general of our troops
was arch-plotter, and that two of the Senators from the
West were fully committed to Burr, shows how ready
much of this element was for unprincipled venture.
We have already seen that the Union was far from
being unified and cemented, even in the older sections.
The French and Creoles and Spaniards of Louisiana
were bitterly hostile to their transfer to our government.   They were ready for a declaration of independence.   Burr had only to get established at New
Orleans to call about him a horde of adventurers.   A
band was known to be in existence in New York City
waiting to join with him as soon as there was a footing.
Notwithstanding Jackson's protest against any attack
on the Union, it is certain that he wrote that all his
troops expected to co-operate with Burr.   Henry
Adams expresses the opinion that " Jackson always
felt sympathy for a duellist who had killed his man ;
but his passions were more exacting in favor of the
man who should drive the Spaniard out of America.
As major-general of the Tennessee Militia, Jackson
looked forward to sharing this exploit."   If not committed to extremes at first, Burr would soon have com-
promised his allies with the States.   In South Carolina
Allston, afterward governor, the richest man in the
South, was his son-in-law and devoted admirer.

Yet Mr. Adams belittles the danger.   He assures
us at one time of Jefferson's reckless neglect of the
warnings sent to him ; at another he says, " No one
denied that if danger ever existed it had passed."

" The President," says Mr. Adams, " continued to
countenance Burr in public, alleging in private that the

people could be trusted to defeat his schemes ; but " he complains, " the people had instituted a government to provide themselves with proper machinery for emergencies ; and the President alone could set it in action." Thus trying to disprove Jefferson's statesmanship, with almost the same breath he tells us that " President Jefferson had already too many feuds on his hands, and Burr had still too many friends, to warrant rousing fresh reprisals, at a time when the difficulties of the administration were extreme."

This is the literal truth, that Mr. Jefferson was forced to move with great caution, and not prematurely, to crush for treason a man who had been, but recently, Vice President, and whose guilt would be scarcely believed without incontrovertible proof. Here was displayed the rare power of Mr. Jefferson to hold his peace and await his time. When at last Burr was arrested, we are again assured by Mr. Adams that " From the first to the last Burr and his counsel never ceased their efforts to convict Jefferson, until the acquittal of Burr began to seem a matter of secondary importance, compared with the President's discomfiture. Over this tournament the chief-justice presided as arbiter." The student of history who goes through the records without bias, will be more than convinced that the Union and the government were in great danger.

Had the plans of Burr succeeded so far as to take possession of New Orleans, it is probable that the Union would have been severed. There would then have followed (1) a recognition of the new government by Tennessee, and possibly Ohio. These States were certainly inclined to believe their interest lay in a Mississippi Confederacy. (2) The navigation of the

Mississippi would have been in the hands of the new government. (3) The recognition by Great Britain would have followed, with an alliance. (4) We should have been forced into our second war with England prematurely, and at odds. (5) New England leaders would have renewed their attempt to create a Northern Confederacy. (6) Three confederacies would probably have taken the place of the single Union ; and the whole history of our continent would have been altered.

APPENDIX TO CHAPTER IV

BURR'S VALEDICTORY TO THE UNITED STATES
SENATE

AS REPORTED BY A SENATOR

On Saturday, the 2d of March, 1805, Mr. Burr took leave of the Senate. This was done at a time when the doors were closed ; the Senate being engaged in executive business, and, of course, there were no spectators. It is, however, said to have been the most dignified, sublime, and impressive that ever was uttered; and the effect which it produced justifies these epithets.

He said he was sensible he must at times have wounded the feelings of individual members. He had ever avoided entering into explanations at the time, because a moment of irritation was not a moment for explanation ; because his position (being in the chair) rendered it impossible to enter into explanations without obvious danger of consequences which might hazard the dignity of the Senate, or prove disagreeable and injurious in more than one point of view ; that he had, therefore, preferred to leave to their reflections his justification ; that, on his part, he had no injuries to complain of; if any had been done or attempted, he was ignorant of the authors ; and if he had ever heard, he had forgotten.

He doubted not but that they had found occasion to observe, that to be prompt was not therefore to be precipitate; and that to act without delay was not always to act without reflection ; that error was often to be preferred to indecision ; that his errors, whatever they might have been, were those of rule and principle, and not of caprice ; that it could not be deemed arrogance in him to say that, in his official conduct, he had known no party—no cause—no friend; that if, in the opinion of any, the discipline which had been established approached to rigor, they would at least admit that it was uniform and indiscriminate.

He further remarked, that the ignorant and unthinking affected to treat as unnecessary and fastidious a rigid attention to rules and decorum ; but he thought nothing trivial which touched, however remotely, the dignity of that body; and he appealed to their experience for the justice of this sentiment, and urged them in language the most impressive, and in a manner the most commanding, to avoid the smallest relaxation of the habits which he had endeavored to inculcate and establish.

But he challenged their attention to considerations more momentous than any which regarded merely their personal honor and character—the preservation of law, of liberty, and the Constitution. "This house," said he, " is a sanctuary; a citadel of law, and of liberty; and it is here—it is here, in this exalted refuge—here if anywhere, will resistance be made to the storms of political frenzy and the silent arts of corruption; and if the Constitution be destined to perish by the sacrilegious hands of the demagogue or the usurper, which God avert, its expiring agonies will be witnessed on this floor."

He then adverted to those affecting sentiments which attended a final separation—a dissolution, perhaps forever, of those associations which he hoped had been mutually satisfactory. He consoled himself, however, and them, with the reflections that, though they separated, they would be engaged in the common cause of disseminating principles of freedom and social order.

He should always regard the proceedings of that body with interest and with solicitude. He should feel for their honor, and the national honor so intimately connected with it ; and took his leave with expressions of personal respect, and with prayers, and wishes, etc.

At the President's on Monday, two of the Senators were relating these circumstances, to a circle which had collected round them. One said that he wished that the tradition might be preserved as one of the most extraordinary events he had ever witnessed. Another Senator being asked, on the day following that on which Mr. Burr took his leave, how long he was speaking, after a moment's pause, said he could form no idea ; it might have been an hour, and it might have been but a moment ; when he came to his senses, he seemed to have awakened as from a kind of trance.

The characteristics of the Vice-President's manner seemed to have been elevation and dignity—a consciousness of superiority, etc. Nothing of that whining adulation ; those canting, hypocritical complaints of want of talents ; assurance of his endeavors to please them ; hopes of their favor, etc. On the contrary, he told them explicitly that he had determined to pursue a conduct which his judgment should approve, and which should secure the suffrage of his own conscience, and that he had never considered who else might be pleased or displeased ; although it was but justice on this occasion to thank them for their deference and respect to his official conduct—the constant and uniform support he had received from every member—for their prompt acquiescence to his decisions ; and to remark, to their honor, that they had never descended to a single motion of passion or embarrassment ; and so far was he from apologizing for his defects, that he told them that, on reviewing the decisions he had had occasion to make, there was no one which, on reflection, he was disposed to vary or retract.

As soon as the Senate could compose themselves sufficiently to choose a president pro tem, they came to the following resolution :

Resolved, unanimously, That the thanks of the Senate be presented to Aaron Burr, in testimony of the impartiality, dignity, and ability with which he has presided over their deliberations, and of their entire approbation of his conduct in the discharge of the arduous and important duties assigned him as president of the Senate ; and that Mr. Smith of Maryland and Mr. White be a committee to wait on him with this resolution.

Attest,   SAM. A. OTIS, Secretary.

To which resolution Colonel Burr returned the following answer to the Senate :

Next to the satisfaction arising from a consciousness of having discharged my duty, is that which is derived from the approbation of those who have been the constant witnesses of my conduct; and the value of this testimony of their esteem is greatly enhanced by the promptitude and unanimity with which it is offered.

I pray you to accept my respectful acknowledgments, and the assurance of my inviolable attachment to the interests and dignity of the Senate.

A. BURR.

JEFFERSON'S MESSAGE ON BURR'S CONSPIRACY

To the Senate and House of Representatives of the United States :

Agreeably to the request of the House of Representatives communicated in their resolution of the 16th instant, I proceed to state, under the reserve therein expressed, information received touching an illegal combination of private individuals against the peace and safety of the union ; and a military expedition planned by them against the territories of a power in amity with the United States ; with the measures I have pursued for suppressing the same.

I had been for some time in the constant expectation of receiving such further information as would have enabled me to lay before the Legislature the termination as well as the beginning and progress of the scene of depravity, so far as it has been enacted on the Ohio and its waters. From this the state of safety of the lower country might have been estimated on probable grounds ; and the delay was indulged the rather because no circumstance had yet made it necessary to call in the aid of the legislative functions. Information now recently communicated, has brought us nearly to the period contemplated. The mass of what I have received in the course of these transactions is voluminous, but little has been given under the sanction of an oath, so as to constitute formal and legal evidence. It is chiefly in the form of letters, often containing such a mixture of rumors, conjectures, and suspicions as renders it difficult to sift out the real facts, and unadvisable to hazard more than general outlines, strengthened by concurrent information or the particular credibility of the relator. In this state of the evidence, delivered sometimes under the restrictions of private confidence, neither safety or justice will permit the exposing names, except that of the principal actor, whose guilt is placed beyond question.

Sometime in the latter part of September I received intimations that designs were in agitation in the western country, unlawful and unfriendly to the peace of the Union, and that the prime mover in these was Aaron Burr, heretofore distinguished by the favor of his country. The grounds of these intimations being inconclusive, the objects uncertain, and the fidelity of that country known to be firm, the only measure taken was to urge the informants to use their best endeavors to get further insight into the designs and proceedings of the suspected persons, and to communicate them to me.

It was not till the latter part of October that the objects of the conspiracy began to be perceived, but still so blended and involved in mystery that nothing distinct could be singled out for pursuit. In the state of uncertainty as to the crime contemplated, the acts done, and the legal course to be pursued, I thought it best to send to the scene where these things were principally in transaction a person in whose integrity, understanding, and discretion entire confidence could be reposed, with instructions to investigate the plots going on, to enter into conference (for which he had sufficient credentials) with the governors and all other officers, civil and military, and with their aid to do on the spot whatever should be necessary to discover the designs of the conspirators, arrest their means, bring their persons to punishment, and to call out the force of the country to suppress any unlawful enterprise in which it should be found they were engaged. By this time it was known that many boats were under preparation, stores of provisions collecting, and an unusual number of suspicious characters in motion on the Ohio and its waters. Besides dispatching the confidential agent to that quarter, orders were at the same time sent to the governors of the Orleans and Mississippi Territories, and to the commanders of the land and naval forces there, to be on their guard against surprise, and in constant readiness to resist any enterprise which might be attempted on the vessels, posts, or other objects under their care ; and on the 8th of November instructions were forwarded to General Wilkinson to hasten an accommodation with the Spanish commander on the Sabine ; and as soon as that was effected, to fall back with his principal force to the hither bank of the Mississippi for the defence of the interesting points on that river.   By a letter received from that officer on the 25th of November, but dated October 21, we learnt that a confidential agent of Aaron Burr had been deputed to him with communications, partly written in cipher and partly oral, explaining his designs, exaggerating his resources, and making such offers of emolument and command to engage him and the army in his unlawful enterprise, as he had flattered himself would be successful.   The General, with the honor of a soldier and fidelity of a good citizen, immediately dispatched a trusty officer to me with the information of what had passed; proceeding to establish such an understanding with the Spanish commandant on the Sabine as permitted him to withdraw his forces across the Mississippi, and to enter on measures for opposing the proposed enterprise.

The general's letter, which came to hand on the 25th of November, as has been mentioned, and some other information received a few days earlier, when brought together, developed Burr's general designs, different parts of which had only been revealed to different informants.   It appeared that he contemplated two distinct objects, which might be carried on either jointly or separately, and either the one or the other first, as circumstances should direct. One of these was the severance of the Union of these States by the Alleghany Mountains ; the other an attack on Mexico. A third object was provided, merely ostensible, to wit, the settlement of a pretended purchase of a tract of land on the Washita claimed by Baron Bastrop. This was to serve as the pretext for all his preparations, an allurement for such followers as really wished to acquire settlements in that country, and a cover under which to retreat, in event of a final discomfiture of both branches of his real design.

He found at once that the attachment of the Western country to the present Union was not to be shaken ; that its dissolution could not be affected with the consent of its inhabitants, and that his resources were inadequate as yet to effect it by force. He took his course then at once ; determined to seize on New Orleans, plunder the bank there, possess himself of the military and naval stores, and proceed on his expedition to Mexico; and to this object all his means and preparations were now directed. He collected from all the quarters where himself or his agent possessed influence, all the ardent, restless, desperate, and disaffected persons who were ready for any enterprise analogous to their characters. He seduced good and well meaning citizens, some by the assurance that he possessed the confidence of the Government, and was acting under its secret patronage, a pretense which secured some credit from the state of our differences with Spain, and others by offers of land in Bastrop's claims on the Washita.

This was the state of my information of his proceedings about the last of November, at which time, therefore, it was first possible to take specific measures to meet them. The proclamation of November 27th, two days after the receipt of General Wilkinson's information, was now issued. Orders were dispatched to every interesting point on the Ohio and the Mississippi, from Pittsburg to New Orleans, for the employment of such force either of the regulars or of the militia, and of such proceedings also of the civil authorities as might enable them to seize on all the boats and stores provided for the enterprise, to arrest the persons concerned, and to suppress effectually the further progress of the enterprise. A little before the receipt of these orders in the State of Ohio, our confidential agent, who had been diligently employed in investigating the conspiracy, had acquired sufficient information to open himself to the governor of that State, and apply for the immediate exertion of the authority and power of the State to crush the combination. Governor Tiffin and the legislature, with a promptitude, and energy, and patriotic zeal which entitle them to a distinguished place in the affections of their sister States, effected the seizure of all the boats, provisions, and other preparations within their reach, and thus gave a first blow, materially disabling the enterprise in its outset.

In Kentucky a premature attempt to bring Burr to justice, without sufficient evidence for his conviction, had produced a popular impression in his favor, and a general disbelief of his guilt. This gave him an unfortunate opportunity of hastening his equipments. The arrival of the proclamation and orders, and the application and information of our confidential agent at length awakened the authorities of that State to the truth, and then produced the same promptitude and energy of which the neighboring State had set the example.   Under an act of their legislature of December 23, militia were instantly ordered to different important points, and measures taken for doing whatever could yet be done.   Some boats (accounts vary from five to double or treble that number) and persons (differently estimated from 100 to 300) had in the meantime passed the Fall of the Ohio, to rendezvous at the mouth of the Cumberland, with others expected down that river.

Not apprised till very late that any boats were building on the Cumberland, the effect of the proclamation had been trusted to for some time in the State of Tennessee ; but on the 19th of December similar communications and instructions with those to the neighboring States were dispatched by express, to the Governor, and a general officer of the western division of the State, and on the 23d of December our confidential agent left Frankfort for Nashville, to put into activity the means of that State also.   But by information received yesterday, I learn that on the 22d of December Mr. Burr descended the Cumberland with two boats merely of accommodation, carrying with him from that State no quota toward his unlawful enterprise.   Whether after the arrival of the proclamation, of the orders, or of our agent, any exertion which could be made by that State, or the orders of the Governor of the State of Kentucky, for calling out the militia at the mouth of the Cumberland, would be in time to arrest these boats and those from the Falls of the Ohio is still doubtful.

On the whole the fugitives from the Ohio, with their associates from the Cumberland, or any other place in that quarter, cannot threaten serious danger to the City of New Orleans.

By the same express of December 19, orders were

sent to the governors of Orleans and Mississippi, supplementary to those which had been given on the 25th of November, to hold the militia of their territories in readiness to co-operate for their defence with the regular troops and armed vessels then under command of General Wilkinson. Great alarm, indeed, was excited at New Orleans by the exaggerated accounts of Mr. Burr, disseminated through his emissaries, of the armies and navies he was to assemble there. General Wilkinson had arrived there himself on the 24th of November, and had immediately put into activity the resources of the place for the purpose of its defence, and on the 10th of December he was joined by his troops from the Sabine. Great zeal was shown by the inhabitants generally; the merchants of the place readily agreeing to the most laudable exertions and sacrifices, for manning the armed vessels with their seamen ; and the other citizens manifesting unequivocal fidelity to the Union, and a spirit of determined resistance to their expected assailants.

Surmises have been hazarded that this enterprise is to receive aid from certain foreign powers ; but these surmises are without proof or probability. The wisdom of the measures, sanctioned by Congress at its last session, has placed us in the paths of peace and justice with the only powers with whom we had any differences, and nothing has happened since which makes it either their interest or ours to pursue another course. No change of measures has taken place on our part ; none ought to take place at this time. With the one, friendly arrangement was then proposed, and the law, deemed necessary on the failure of that, was suspended to give time for a fair trial of the issue. With the same power friendly arrangement is now proceeding under good expectations, and the same law deemed necessary on the failure of that is still suspended, to give time for a fair trial of the issue. With the other, negotiation was in a like manner then preferred, and provisional measures only taken to meet the event of rupture. With the same power negotiation is still preferred, and provisional measures only are necessary to meet the event of rupture. While therefore we do not deflect in the slightest degree from the course we then assumed and are still pursuing with mutual consent to restore a good understanding, we are not to impute to them practices as irreconcilable to interest as to good faith, and changing necessarily the relations of peace and justice between us to those of war. These surmises are therefore to be imputed to the vaunting of the author of this enterprise, to multiply his partisans by magnifying the belief of his prospects and support.

By letters from General Wilkinson of the 14th and 18th of December which came to hand two days after the date of the resolution of the House of Representatives, that is to say on the morning of the 18th inst., I received the important affidavit, a copy of which I now communicate, with extracts of so much of the letters as comes within the scope of the resolution. By these it will be seen that of three of the principal emissaries of Mr. Burr whom the General had caused to be apprehended, one had been liberated by habeas corpus, and two others, being those particularly employed in the endeavor to corrupt the General and army of the United States, have been embarked by him for ports in the Atlantic States, probably on the consideration that an impartial trial could not be expected during the present agitation of New Orleans, and that that city was not as yet a safe place of confinement. As soon as these persons shall arrive they will be delivered to the custody of the law, and left to such course of trial, both as to place and process, as its functionaries may direct. The presence of the highest judicial authorities, to be assembled at this place within a few days, the means of pursuing a sounder course of proceedings here than elsewhere, all the aid of the executive means, should the judges have occasion to use them, render it equally desirable for the criminals as for the public that, being already removed from the place where they were first apprehended, the first regular arrest should take place here, and the course of proceedings receive here its proper direction.

Signed,   THOMAS JEFFERSON.

AARON BURR AT HIS TRIAL

Blennerhasset writes in his journal September 13, 1807, " I visited Burr this morning. He is as gay as usual, and as busy in speculations on reorganizing his projects for action, as if he had never suffered the least interruption. He observed to Major Smith and me that in six months our schemes could all be remounted ; that we could now new model them in a better mould than formerly, having a better view of the ground, and a more perfect knowledge of our men. We were silent. It should yet be granted that, if Burr possessed sensibility of the right sort, with one-hundredth part of the energies for which with many he has obtained such ill-grounded credit, his first and last determination with the morning and the night should be the destruction of those enemies who have so long and so cruelly wreaked their malicious vengeance upon him." September 16, 1807, " I was glad to find Burr had at

last thought of asking us to dine with him. The dinner was neat, and followed by three or four sorts of
wine. Splendid poverty ! During the chit chat after
the cloth was removed, a letter was handed to Burr,
next to whom I sat. I immediately smelt musk. Burr
broke the seal, put the cover to his nose, and then
handed it to me saying, ' This amounts to a disclos-
ure.' I smelt the paper and said, ' I think so.' The
whole physiognomy of the man now assumed an alteration and vivacity that to a stranger who had never seen
him before would have sunk full fifteen years of his
age. ' This,' said he, ' reminds me of a detection once
very nearly practiced on me at New York. One day a
lady stepped into my library while I was reading, and
giving me a slap on the cheek said, Come, tell me
directly what little French girl, pray, have you had
here ? The abruptness of the question and surprise
left me little doubt that the discovery had been completely made. So I thought it best to confess the whole
fact ; upon which the inquisitress burst into a loud
laugh on the success of her artifice, which she was led
to play off from the mere circumstance of having smelt
musk in the room.' After some time Martin and Pre-
vost withdrew, and we passed to the topic of our late
adventures on the Mississippi, in which Burr said little,
but declared he did not know of any reason to blame
General Jackson of Tennessee for anything he had done
or omitted. But he declares he will not lose a day after
the favorable issue at the capital, of which he has no
doubt, to direct his entire attention to setting up his
projects which have only been suspended, on a better
model, in which work he says he has even made here
some progress." October 8, 1807. " I called on
Burr this morning, when he at last mentioned to me

that he was preparing to go to England, and he wished
to know whether I could give him letters. I answered
that I supposed he meant London, as his business
would probably be with people in office ; that I knew
none of the present ministry. He replied that he
would be glad to get letters to anyone. We can only
conjecture his design. For my part I am disposed to
suspect he has no serious intent of reviving any of his
speculations in America, or even of returning from
Europe if he can get there.''

TESTIMONY OF WILLIAM EATON AT THE TRIAL OF AARON BURR

Mr. Eaton. During the winter of 1805, 6, at the
city of Washington, Aaron Burr signified to me, that
he was organizing a military expedition to be moved
against the Spanish provinces, on the southwestern
frontiers of the United States : I understood under the
authority of the general government. From our existing controversies with Spain, and from the tenor of the
President's communications to both houses of Congress, a conclusion was naturally drawn, that war with
that power was inevitable. I had just then returned
from the coast of Africa, and having been for many
years employed on your frontier, or a coast more barbarous and obscure, I was ignorant of the estimation in
which Colonel Burr was held by his country. The
distinguished rank he held in society, and the strong
marks of confidence which he had received from his
fellow citizens, did not permit me to doubt of his patriotism. As a military character, I had been made
acquainted with none within the United States, under
whose direction a soldier might with greater security

confide his honor than Colonel Burr.   In case of my
country's being involved in a war, I should have thought
it my duty to obey so honorable a call, as was proposed
to me.   Under impressions like these, I did engage to
embark myself in the enterprise, and pledged myself to
Colonel Burr's confidence.   At several interviews, it
appeared to be his intention to convince me by maps
and other documents, of the feasibility of penetrating
to Mexico.   At length, from certain indistinct expressions and innuendoes, I admitted a suspicion that
Colonel Burr had other projects.   He used strong
expressions of reproach against the administration of
the government ; accused them of want of character,
want of energy, and want of gratitude.   He seemed
desirous of irritating my resentment by dilating on certain injurious strictures I had received on the floor of
Congress, on account of certain transactions on the
coast of Tripoli ; and also on the delays in adjusting
my accounts for advances of money, on account of the
United States ; and talked of pointing out to me modes
of honorable indemnity.   I will not conceal here, that
Colonel Burr had good reasons for supposing me disaffected towards the government ; I had indeed suffered
much, from delays in adjusting my accounts for cash
advanced to the government, whilst I was consul at
Tunis, and for the expense of supporting the war with
Tripoli.   I had but a short time before been compelled
ingloriously to strike the flag of my country, on the
ramparts of a defeated enemy, where it had flown for
forty-five days.   I had been compelled to abandon my
comrades in war, on the fields where they had fought
our battles.   I had seen cash offered to the half-
vanquished chief of Tripoli, (as he himself had acknowledged,) as the consideration of pacification.

Mr. Wickham.   By whom?

Answer. By our negotiator, when as yet no exertion
had been made by our naval squadron to coerce that
enemy.   I had seen the conduct of the author of these
blemishes on our then proud national character, if not
commended—not censured ; whilst my own inadequate
efforts to support that character were attempted to be
thrown into shade.   To feelings naturally arising out
of circumstances like these, I did give strong expression.   Here I beg leave to observe, in justice to myself,
that however strong these expressions, however harsh
the language I employed, they would not justify the
inference, that I was preparing to dip my sabre in the
blood of my countrymen ; much less of their children,
which I believe would have been the case, had this
conspiracy been carried into effect.

Mr. Martin objected to this language.

I listened to Colonel Burr's mode of indemnity ; and
as I had by this time begun to suspect that the military
expedition he had on foot was unlawful, I permitted
him to believe myself resigned to his influence, that I
might understand the extent and motive of his arrangements.   Colonel **Burr** now laid open his project of
revolutionizing the territory west of the Alleghany ;
establishing an independent empire there ; New Orleans to be the capital, and he himself to be the chief ;
organizing a military force on the waters of the Mississippi, and carrying conquest to Mexico.   After much
conversation, which I do not particularly recollect,
respecting the feasibility of the project, as was natural,
I stated impediments to his operations ; such as the
republican habits of the citizens of that country, their
attachment to the present administration of the government, the want of funds, the opposition he would expe-

rience from the regular army of the United States, stationed on the frontier ; and the resistance to be expected from Miranda, in case he should succeed in republicanizing the Mexicans. Colonel Burr appeared to have no difficulty in removing these obstacles. He stated to me, that he had in person, (I think the preceding season,) made a tour through that country ; that he had secured to his interests and attached to his person, (I do not recollect the exact expression, but the meaning, and I believe, the words were,) the most distinguished citizens of Tennessee, Kentucky, and the territory of Orleans ; that he had inexhaustible resources and funds ; that the army of the United States would act with him ; that it would be reinforced by ten or twelve thousand men from the above mentioned States and Territories ; that he had powerful agents in the Spanish territory, and " as for Miranda," said Mr. Burr facetiously, " we must hang Miranda." In the course of several conversations on this subject, he proposed to give me a distinguished command in his army ; I understood him to say, the second command. I asked him who would command in chief. He said, General Wilkinson. I observed, that it was singular, he should count upon General Wilkinson ; the distinguished command and high trust he held under government, as the commander-in-chief of our army, and as governor of a province, he would not be apt to put at hazard for any prospect of precarious aggrandisement. Colonel Burr stated, that General Wilkinson balanced in the confidence of his country ; that it was doubtful whether he would much longer retain the distinction and confidence he now enjoyed ; and that he was prepared to secure to himself a permanency. I asked Colonel Burr, if he knew General Wilkinson. He said, yes ; and echoed

the question. I told him that twelve years ago I was
at the same time a captain in the wing of the legion of
the United States, which General Wilkinson commanded, his acting brigade-major, and aid-de-camp ;
and that I thought I knew him well. He asked me,
what I knew of General Wilkinson ? I said, I knew
General Wilkinson would act as lieutenant to no man
in existence. " You are in an error," said Mr. Burr,
" Wilkinson will act as lieutenant to me." From the
tenor of much conversation on this subject I was prevailed on to believe that the plan of revolution meditated by Colonel Burr, and communicated to me, had
been concerted with General Wilkinson, and would
have his co-operation ; for Colonel Burr repeatedly,
and very confidently expressed his belief, that the
influence of General Wilkinson with his army, the
promise of double pay, and rations, the ambition of his
officers, and the prospect of plunder and military
achievements, would bring the army generally into the
measure. I pass over here, a conversation which took
place between Colonel Burr and myself, respecting a
central revolution, as it is decided to be irrelevant, by
the opinion of the bench.

Mr. Hay.—You allude to a revolution for overthrow-
ing the government at Washington, and of revolutionizing the eastern States.

I was passing over that, to come down to the period
when I supposed he had relinquished that design, and
adhered to the project of revolutionizing the west.

Mr. Wickham.—What project do you mean ?

Answer. A central general revolution. I was
thoroughly convinced myself, that such a project was
already so far organized, as to be dangerous, and that
it would require an effort to suppress it. For in addi-

tion to positive assurances that Colonel Burr had of assistance and cooperation, he said, that the vast extent of territory of the United States, west of the Alleghany Mountains, when offered to adventurers with a view on the mines of Mexico, would bring volunteers to his standard from all quarters of the Union.   The situation which these communications, and the impressions they made upon me, placed me in, was peculiarly delicate. I had no overt act to produce against Colonel Burr. He had given me nothing upon paper ; nor did I know of any person in the vicinity, who had received similar communications, and whose testimony might support mine.   He had mentioned to me no persons as principally and decidedly engaged with him, but General Wilkinson ; a Mr. Allston, who, I afterwards learned, was his son-in-law ; and a Mr. Ephraim Kibby, who I learnt was late a captain of rangers in Wayne's army. Of General Wilkinson, Burr said much, as I have stated; of Mr. Allston, very little, but enough to satisfy me that he was engaged in the project ; and of Kibby, he said, that he was brigade-major in the vicinity of Cincinnati (whether Cincinnati in Ohio or in Kentucky, I know not,) who had much influence with the militia, and had already engaged the majority of the brigade to which he belonged, who were ready to march at Mr. Burr's signal.   Mr. Burr talked of this revolution as a matter of right, inherent in the people, and constitutional ; a revolution which would rather be advantageous than detrimental to the Atlantic States ; a revolution which must eventually take place ; and for the operation of which, the present crisis was peculiarly favorable. He said there was no energy to be dreaded in the general government, and his conversations denoted a confidence that his arrangements were so well made, that he should meet with no opposition at New Orleans ; for the army and chief citizens of that place were now ready to receive him. On the solitary ground upon which I stood, I was at a loss how to conduct myself, though at no loss as respected my duty. I durst not place my lonely testimony in the balance against the weight of Colonel Burr's character ; for by turning the tables upon me, which I thought any man, capable of such a project, was very capable of doing, I should sink under the weight. I resolved therefore with myself, to obtain the removal of Mr. Burr from this country in a way honorable to him ; and on this I did consult him, without his knowing my motive. Accordingly, I waited on the President of the United States, and after a desultory conversation, in which I aimed to draw his view to the westward, I took the liberty of suggesting to the President, that I thought Colonel Burr ought to be removed from the country, because I considered him dangerous in it. The President asked where we should send him ? Other places might have been mentioned, but I believe that Paris, London, and Madrid, were the places which were particularly named. The President without positive expression (in such a matter of delicacy) signified that the trust was too important, and expressed something like doubt about the integrity of Mr. Burr. I frankly told the President, that perhaps no person had stronger grounds to suspect that integrity than I had ; but that I believed his pride of ambition had so predominated over his other passions, that when placed on an eminence, and put on his honor, a respect to himself would secure his fidelity. I perceived that the subject was disagreeable to the President, and to bring him to my point in the shortest mode, and at the same time, point to the danger, I said to him that I expected that we should in eighteen months have an insurrection, if not a revolution, on the waters of the Mississippi. The President said he had too much confidence in the information, the integrity, and attachment to the Union of the citizens of that country to admit any apprehensions of that kind. The circumstance of no interrogatories being made to me, I thought imposed silence upon me at that time and place. Here, sir, I beg indulgence to declare my motives for recommending that gentleman to a foreign mission at that time ; and in the solemnity with which I stand here, I declare that Colonel Burr was neutral in my feelings ; that it was through no attachment to him that I made that suggestion, but to avert a great national calamity which I saw approaching ; to arrest a tempest which seemed lowering in the west ; and to divert into a channel of usefulness those consummate talents, which were to mount " the whirlwind and direct the storm." These, and these only, were my reasons for making that recommendation.

About the time of my having waited on the President, or a little before, (I cannot however be positive whether before or after) I determined at all events to have some evidence of the integrity of my intentions, and fortify myself by the advice of two gentlemen, members of the House of Representatives, whose friendship and confidence I had the honor long to retain, and in whose wisdom and integrity, I had the utmost faith and reliance. I am at liberty to give their names if required. I do not distinctly recollect, but I believe, that I had a conversation with a Senator on the subject. I developed to them all Mr. Burr's plans. They did not seem much alarmed.

An interesting interview with Burr is given by
William H. Seward in his Autobiographic Reminiscences. '' Burr had returned from his long exile and
disgrace in Europe, and resumed the practice of law in
New York. He appeared at Albany, and by a courte-
ous note applied for an interview, which of course I
could not refuse. He opened the interview with expressions of sympathy in my political opinions, and
then easily digressed into reminiscences of the war.
Of the disastrous attack upon Quebec, of the battle of
Monmouth, of the military family of Washington, of
his generals', Greene, Gates, Lafayette ; of Talleyrand ;
of Dr. Franklin; and even of his great rival Hamilton
whom he had slain. The interview was held in my
family on a Sunday. He suffered no passage in it to
occur without addressing some pleasing compliment to
my wife, and all the while held one or both of my children on his knee. At last he came to the object of his
visit. I thought I was wary as well as firm, in declin-
ing his request that I would facilitate his application to
reinstate a chancery suit. . . . His conversation
was fascinating, and in one sense instructive, though
on most subjects prejudiced and insincere. He represented Washington as being entirely without independence of character, and without talent, and completely under the influence of Alexander Hamilton.
Burr said that Washington did not trust himself to write
a billet of invitation, or acceptance of a dinner ; and
therefore employed Hamilton to do it. He said Washington was formal, cold, and haughty. On the other
hand he especially admired Franklin ; whom he represented as all suavity, courtesy, and kindness. He

described him as more eminent in his time as a genial
wit and humorist in the social circle, than as a philosopher ; and he placed Franklin always in the same
category with Talleyrand.   While he conceded to
Hamilton great talent, he represented him as a parasite
of Washington, unamiable and ungenerous toward all
others.   When I referred to the histories of the Revolution, and especially to Marshall's Life of Washington, as differing from his own representations, he
replied that the histories were all partial, interested,
' unreliable, and false.'   ' I was myself present,' said he,
at a skirmish which he had with the enemy at Monmouth, N. J.   Of course I well knew what occurred
there.   I have read accounts of that battle from a
dozen different histories, and if it were not that the
date of the battle and the place where it was fought
were mentioned, I should not recognize in the description that it was the battle of Monmouth at all.'   He
was severely satirical upon Jefferson, who he said he
verily believed would have run away from Monticello,
if he had heard that he (Burr) had approached as near
as Alexandria or Georgetown.''

CHAPTER V

WE must turn back to New England far more quickly than is agreeable to meet the next attempt to break up the Union. The election of Jefferson had been a shock to conservatism in both Church and State ; but as we have seen, opposition gave way before the splendid success and moderation of his administration. Unfortunately our foreign relations during his second term grew more and more strained. The larger part of the stipulations of the Jay treaty terminated in 1804. England had however overreached even that unjust agreement. Our neutral commerce was held open to seizure and confiscation. The right was asserted by her war vessels to overhaul our ships and impress seamen. If we had secured independence, England meant to make it worth to us as little as possible. The king sent out a proclamation affirming the right of Great Britain to secure her own seamen wherever found. Pickering, always careful to be on the wrong side, defended the British claim. Adams insisted that the doctrine was novel and unjustifiable. "It is," he said, "in direct contradiction and violation of every principle of English liberty. Is there any reason why another proclamation should not appear commanding all the officers of the English army in Canada to go over the line, and take by force all the king's subjects they can find in our villages?" The principle was precisely the same ; but many Federals were willing to submit to the insolence on the ocean.

Meanwhile the struggle between France and England had been renewed on the seas. It was necessary for each one to secure the alliance of the United States, or to compel subservience. In 1806 England declared all the French ports blockaded. France retorted with a similar fulmination. It was all a matter of paper ; nevertheless it subjected neutral ships to danger from both parties. Our shippers for the most part favored England, believing her to be the master of the ocean. The government looking at the justice of the case, felt that France was at least the less unjust of the two. But France was obnoxious in another direction. Our boundary lines with Florida, were under dispute ; and Talleyrand would not permit Spain to accede to our demands. For a time there was about equal danger of war in three directions. There was a short period when Jefferson was almost persuaded to seek English alliance in a war against Spain and France. Monroe negotiated a new treaty with Great Britain ; but it was a stingy affair, yielding no material point to the United States ; and especially ignoring the question of impressment. It was rejected at once by our executive, without submitting it to Congress. English war vessels coasted along our shores to overhaul our ships ; and even entered our harbors in chase of their prey. The Chesapeake, sailing from Hampton Roads in August of 1807, was overhauled by the British man-of-war Leopard ; and not yielding to the orders of the admiral, was fired upon until the American flag was lowered. Embargo

was recommended by the President as a temporary expedient.   It was a sacrifice of a nature so serious that
it could be excusable only in a case like that in which
the Union was then placed.   It had this advantage that
it was neither tame submission, nor was it to proclaim
a war, which would be fought hopelessly on our part.
It was simply a resolution to withdraw from the trade
of the nations.   It was not much unlike the scheme so
long pursued by China and Japan, of shutting their
ports against the world.   Nominally to ensure the
safety of our ships from capture, it was intended to
make the inhabitants of Great Britain and France suffer
for lack of goods, only to be secured from America.
But while this would be endurable for an agricultural
race, it could create only misery for a commercial
people.   The embargo was specially injurious to New
England shipping.   Acquiescing for a time under the
lead of the Adams, both father and son, Joseph Story,
Eustis, and others, those States soon grew restive.
For some years past New England had become less and
less agricultural ; and more and more manufacturing
and commercial.   It could not eat its goods ; it must
ship them to market.   Its ships were rotting at the
wharves.

The almost suppressed leaders of 1804 came from
their retreat.   Pickering denounced the embargo as a
first step to war with England.   The Massachusetts
legislature, which at first endorsed the measure, in its
next session condemned it.   1808 brought the election
of Madison ; a triumph of Republicanism in spite of
growing disaffection with the embargo.   In the early
part of 1809 a Force Bill was passed by Congress, which
allowed the use of the army and navy in enforcing the
embargo, and making seizures.   In the Boston papers

this act was printed in mourning. Public meetings
were held memorializing the legislature. The legislature took strong ground in the way of justifying Great
Britain, and demanding of Congress the repeal of the
embargo, and a declaration of war at once with France.
Pickering raged with great heat ; but Cabot managed
to suppress his most violent fulminations. The Massachusetts legislature adopted a report intensely British,
and declaring the Enforcement Act " not legally binding." Resistance was recommended. It was nullification as strong as that of 1798. Preachers preached
it ; politicians harangued it. Connecticut was not
one whit behind her neighbor. John Quincy Adams
vouches that the Essex Junto planned a New England
convention to consider secession. DeWitt Clinton,
openly charging this in the New York legislature,
secured the passage of strong resolutions at Albany to
sustain the national policy. The New York Federalists
retorted that the State was the victim of the Clintons,
who ruled as political bosses. They issued an Address
from Albany in which they said : "Bonaparte considered the embargo as evidence of our submission to
his will, and there is too much reason to believe it was
so intended by our government. Yet when the freemen of New England are unwilling to be manacled as
slaves, the legislature of this State denounces their
resolutions as seditions."   Adams was so outspoken
in opposition to plotters that the Massachusetts legislature passed a vote of censure ; and he resigned as
Senator from that State. But New England was really
suffering ; and this time it was not the Federalist leaders only who created the revolt. The people peppered
Washington with protests. Of the five New England
States Mr. Madison carried only Vermont in 1808.

The embargo was not endurable. It was raised in March of 1809. A Non-Intercourse Act was substituted; putting both France and Great Britain under the same ban; while allowing trade with other nations.

It was evident that war was inevitable, and must be prepared for. England, who had promised reparation for the Chesapeake outrage, withdrew her promise. The British Cabinet grew more unyielding and insulting. France was less unfriendly; England exasperating. Castlereagh in 1811 allowed that the English navy had made sixteen hundred seizures of undoubted American sailors. The United States, with the exception of New England, demanded war. At Boston in caucus, resolutions of bitter condemnation of the administration were carried; and others declaring for peace. But in Washington the war drift was assured. A resolution to empower the President to call out 100,000 militia, to serve for six months, was debated. Calhoun, hardly more than a boy, handsome and captivating, made his maiden speech. He said, "The question even in the opinion of our opponents, is reduced to this single point. What shall we do; abandon, or defend, our own commercial and maritime rights; and the personal liberties of our citizens exercising them? These rights are attacked; and war is the only means of redress. Negotiation has been resorted to till it is hopeless. The evil still grows. Sir! which alternative this House ought to sustain is not for me to say." Clay invoked the patriotism of the people to sustain the honor of the nation.

England insisted on continuing her decree of a blockade of France; to which Napoleon responded with the famous Berlin and Milan Decrees, aimed at all neutral trade with England. In return England retaliated by Orders in Council. By these Orders the United States could only trade with France by passing first through English ports, and paying port duties. If a cargo was taken home again, the carrying vessel must go through a British port, and pay another duty. The Milan Decree of Napoleon declared that if any neutral vessel did pay a port tax to England it should be lawful to seize it, and that it should be seized. To America's protest he replied, " A vessel that has paid an English tax is denationalized. I have heretofore negotiated with America free." He would not consider our rights so long as we were " slaves to England's policy." The Berlin Decree was also a part of what Napoleon called " The Continental System." It was simply the highest kind of " protection," at the expense of English commerce. The emperor could not whip England with arms ; he proposed to try commercial warfare. The battle of Friedland put all Europe under his will. In 1807 by the treaty of Tilsit he induced Russia to join in with this system. He ordered Portugal and Denmark to close their ports against English commerce. " Great Britain," he said, " shall be destroyed. I will permit no nation to receive a minister from Great Britain until she shall have renounced her maritime tyranny." Neutrality on the ocean must cease. It was a magnificent stroke of brute-force policy ; but to accomplish it destroyed Napoleon. Between the two colliding forces the United States suffered beyond endurance. It was left to Jefferson and Madison to choose which party to attack. Could anything be accomplished by attacking either ? The Federalists were heart and pen with Great Britain. The Republicans relied on diplomacy. Jefferson wrote " I have ever been anxious to avoid a war with England. I hoped we could coerce her to justice by peaceable means."

Napoleon offered to abrogate his decrees as soon as England would revoke her blockade. The United States would not yield the point that any nation, by a mere declaration that half of civilization was under blockade, could therefore seize and confiscate our merchant vessels. By England we were prevented from trading with the continent of Europe. France in return forbade trade with England or the carrying of English goods. This involving of neutral nations in their selfish contests was ruinous to our commerce. Hundreds of American vessels were seized and sold. Our government insisted on freedom of trade. Mr. Madison demanded that England should respond favorably to the French offer, and annul her paper blockades. In a message of June 1st, 1812, he said to Congress, " We behold in fine on the side of Great Britain a state of war against the United States, and on the side of the United States a state of peace toward Great Britain."

Not only was the blockade sustained but Orders in Council of the British government prohibited trading by merchants anywhere that England herself was not permitted to trade. So that whatever suffering she involved herself in, she compelled other sovereign powers to share with her. Most outrageous of all was the Order in Council declaring every country, anywhere, from which the British flag was excluded, to be practically under blockade. " All ports and places of France, and her allies, or of any other country at war with his Majesty, and all other ports or places in Europe from which, although not at war with his Majesty, the British flag is excluded, and all ports in the colonies of his Majesty's enemies, shall from henceforth be subject to the same restrictions in point of trade and navigation as if the same were actually blockaded in the most strict and vigorous manner." The object of France was to impoverish England; the object of England to tax the trade of the world to keep herself equal to the struggle. She intended by force of her navy to make her harbors " the warehouse of the world's commerce." It was a struggle to the death of the two leading powers of Europe. Napoleon wrote, " Either our government must destroy the English monarchy, or must expect itself to be destroyed." In the end England was nearly ruined ; and Napoleon was wrecked. Meanwhile all the world was impoverished ; and half the honest commerce of civilization turned to smuggling. The English navy vessels were practically only so many pirates. Degeneration marked every phase of civilized society.

Madison justly classed these Orders in Council as a declaration of war on the United States. He called attention also to the fact that secret agents had been commissioned to travel through New England to find out what disaffection existed, and to foment it. The commission of one John Henry was made out by Sir James Craig, Governor of the British provinces of North America. This commission asked for " the earliest information as to how far, in case of war, England could look for assistance." Henry reported that "the Governor of Vermont made no secret of his determination, as commander-in-chief of the militia, to refuse obedience to any command of the general government." From Boston he wrote in a similar vein. But not conceiving himself well treated, this spy turned all his papers over to our government. Madison declared it was perfidy quite unendurable on the part of a neighboring government. The committee on foreign relations reported that the British government had been deliberately pressing measures to divide the States.

Remembering the course of Mr. Merry, and that of the British cabinet in corresponding with Mr. Burr, in 1804, and the vigorous efforts to break up the United States that had characterized the course of Great Britain up to that very recent period, it is probable that the exposure of the mission of Mr. Henry in New England did not uncover more than a tithe of the treachery that was undertaken.

In April of 1812 Congress practically declared war by authorizing the President to call on the State executives to organize their militia for marching at a moment's warning. A formal declaration followed two months later. This act of the government called out at once from the Federalists in Congress an address to the people of New England declaring the war needless and unwise. It bore prompt fruit. The Massachusetts House of Representatives voted an address denouncing the war as a wanton sacrifice of the interests of New England. This address called for town meetings, to consult as to the best methods for protest and action—; not to aid the government but to hinder it. "Express your sentiments without fear" was the advice given. "And let the sound of your disapprobation of this war be loud and deep. If your sons be torn away from you by conscription, consign them to the care of God ; but let there be no *volunteers* except for defensive war." This was at the very outset practical secession. The State of Massachusetts asserted its supreme right, inside the Union, to decide above the Nation, and for the nation. It furthermore refused to fight in any war where not directly assailed on its own soil.

The appointments to command both in the East and

the West were unfortunately bad. Scott said, "They consisted generally of swaggerers, dependents, decayed gentlemen, and others—fit for nothing else." Hull was by all odds the better selection, but his surrender of Detroit was held by the country to be the result of rank cowardice. This was unjust to Hull, whose force was wholly inadequate to the work in hand. The truth was he was a Massachusetts gentleman put in command of Western frontiersmen, and Indian fighters ; and while he did not know how to command such men, they did not know how to obey him. Dearborn who had been appointed to command in the East, stayed in Boston waiting for something to turn up. Every possible hindrance was thrown in the way of his securing enlistments. Those who did enlist were arrested on real or fictitious charges of debt ; and the Courts cheerfully insisted that " while the man was a debtor he was the property of the creditor, and could not leave the State " if he would.

Dearborn at no time seems to have thought of the need of sustaining Hull, or of being sustained by him. He was still at Boston grumbling at Massachusetts treason, while Hull, obedient to orders, urging him on, crossed the Detroit River with a small army. The whole British force was able unhindered to face him, and he was quickly back in Detroit, with the English and Indians on his heels. Detroit was taken instead of Canada. Hull had not misunderstood the situation ; but had written to Washington that he hoped there was a force operating at Niagara to keep the British from concentrating on him and his command. Dearborn had not yet got out of Massachusetts when Hull surrendered. He it was who deserved court-martial and cashiering. Henry Adams aptly shows the

14

wretched incompetency of the Secretary of War when
he says, " The government expected General Hull
with a force, which it knew did not at the outset exceed 2000 effectives, to march 200 miles, constructing a
road as he went ; to garrison Detroit ; to guard at least
60 miles of road under the enemy's guns ; to face a
force at least equal to his own ; with another savage
force of unknown numbers in his rear ; to sweep the
Canadian peninsula of British troops ; to capture the
fortress at Malden and the British fleet on Lake Erie ;
and to do all this without the aid of a fort between
Sandusky and Quebec."

In July peremptory orders came to Dearborn to start
at once for Albany, gather there all the recruits he was
able to secure from New England and New York, and
then, making Niagara the base of operations, invade
Canada.   The time had now come for the Federalists
to take decisive action.   They did not intend to fight.

Governor Griswold of Connecticut professed that he
did not believe the militia of that State could be ordered
to obey a Continental officer.   Months were spent in
disputing about the rank of officers of the militia.
Dwight, the apologist of these scandalous measures,
sums up the position.   " If the New England States
had given up their militia at the requisition of the
President of the United States, and in a total disregard
of the Federal Constitution, a precedent would have
been established that might, and one day or other in
all probability would, have proved fatal to the liberties
of the country.   By the act of Congress the President
was authorized at his own discretion to call into the
public service one hundred thousand militia.   He was
constituted sole judge of the time when they should be
ordered into the field, and of the numbers that should

be called for on any given occasion. By depriving them of their constitutional right to be commanded by their own officers, and placing them under officers of the regular army, they might be pent up in garrisons, or be sent to any distant point of military operations which the President should designate. In this way the several States would have been stripped of their natural and constitutional defenders, and left exposed to all the evils such a condition presupposes ; while the militia themselves would be subjected to all the hardships and degradation which are experienced in standing armies."

In other words, the New England States undertook to decide against Congress ; to thwart the general government ; and in time of war. To confirm themselves in this nullification schedule they soon began to complain of the method in which the war was carried on. The Supreme Court of Massachusetts affirmed that the States had a right to decide whether exigencies existed that warranted the calling forth of the militia. This meant that any governor could nullify a declaration of war by the national government, and could refuse to furnish troops. Dearborn agreed that the militia of each State should be commanded by majors from the State furnishing them. Governor Griswold promptly shifted his treason to another basis of complaint. Legislatures were called together to deliberate. Questions were referred to State courts. Every possible delay was interjected. The legislature of Connecticut resolved " that the conduct of his excellency the governor in refusing to order the militia of this State into the service of the United States on the requisition of the Secretary of War meets with the entire approbation of this Assembly."

President Madison in his message, November 4th,

1812, said of the action " of the governors of Massachusetts and Connecticut": " It is obvious that if the authority of the United States to call into service and command the militia, for the public defense, can be thus frustrated, even in a state of declared war, they are not one nation for the purpose most requiring it ; and that the public safety may have no other resource than in those large and permanent military establishments which are forbidden by the principles of free government." Meanwhile complaints were heard that the same government, that " had no right to call for the militia of New England to go outside that section," was neglecting to defend New England against invasion. In other words the general government must furnish assistance to guard every mile of New England coast, while the latter refused to send one of her militia outside State borders.

The South and West were overwhelmingly loyal. Those portions of the country which had been most recently erected into States or Territories were rapidly filling with people who were devoted to the Union. The Burr episode had been of vast benefit in teaching caution to those who had been too ready for new adventures, and even filibustering exploits. Clay had filled Kentucky with enthusiasm. Tennessee followed Jackson, who was commanding in the Southwest, with equal zeal. These two States were conspicuous not only for loyalty but for the excellence of their troops, and the bravery of their officers.

While Hull was surrendering at Detroit, Dearborn was waking up. Van Rensselaer, one of his subordinates, in October of 1812 crossed into Canada to capture Queenstown. He got 600 troops over the river ; and the rest, who were militia, refused to leave the States. They " were enlisted to protect their State, not to invade foreign territory." For once the spirit of Federalism tainted the New York volunteers. It was however a solitary case, and due as much to the officers as the men. The 600, including Captain Wool- worth, Lieutenant-Colonel Scott, and Brigadier-General Wadsworth were made prisoners by the English. Van Rensselaer resigned, and was succeeded by a light- headed Irishman with no end of eloquence. His pro- clamations and addresses were copied after Napoleon's —with an added touch of Patrick Henry. He bade the mutinous militia " Think on your country's honors torn ; her rights trampled on ! her sons enslaved ! her infants perishing by the hatchet ! Be strong ! Be brave ! and let the ruffian power of the British king cease on this continent." Right or wrong, this valiant fellow made a flat back-out of his promised invasion ; and his troops, believing him to be a coward, tried to kill him.

The Federalists took advantage of disaster to press their cause all the harder. It was declared to be capable of proof that Madison (and Jefferson with him) were in league with Napoleon. That it was on this account that Napoleon had sold Louisiana ; that it was intended to aid Napoleon in subduing all Europe, and England as well, to his regime and authority. The clergy preached and the politicians orated in the same strain. " The Best " rapidly regained power over the democratic masses. A Presidential election must go on in the very crisis of the war. New England except Vermont cast its votes for De Witt Clinton. But in spite of all antagonism, Madison carried Pennsylvania and the rest of the States, except New Jersey and Delaware and New York with a part of the Maryland

electors. Massachusetts Federalists had an overwhelming majority of twenty-four thousand. "The
Peace Party" as they dubbed themselves was growing
stronger in Congress. The terrible embarrassments of
the administration were not limited to disasters in
battle, and the opposition of faction ; the Treasury was
bankrupt. New England banks refused aid. There
was jealousy all around. New York was not cordial
toward the Virginians ; but it was loyal. Armstrong,
the Secretary of War, who was considered the great
rival of Monroe for future political preference, was a
New Yorker ; and their not over cordial relations were
felt both in the cabinet and at the seat of war.

1812 drew toward a close, and the condition of affairs
was pitiful. Madison in his Message spoke warmly of
the course taken by New England, as practically destroying the union as Washington conceived it ; and as
the Federalists above all had insisted on creating it.
Instead of one nation we were acting as two, in the
face of the enemy. He defined the war as an expression of our determination to compel England to formally
renounce the right of impressment of sailors from
American ships. The Treasury it was announced must
borrow twenty millions of dollars for the next year's
campaign. Gallatin, the very able Secretary of the
Treasury, expressed his conviction that no such loan
could be secured ; but that the money must be raised
by taxation. New England had so far turned its energies to manufactures that it was beginning to care less
for commerce. It was now approaching a readiness to
see the embargo on foreign trade made more stringent
and even permanent, under the title of a protective
tariff.

The campaign of 1813 somewhat retrieved our dis-

grace.   Harrison in command at the West chased
Proctor out of Michigan, and whipped him in Canada.
Our naval fights continued to be gallantly conducted
and successful.   While Brown fought well at Sackett's
Harbor ; Dearborn proved to be such a totally incompetent man that he was removed.   But Wilkinson, who
had some months previously been ordered to repair
from New Orleans to the Niagara frontier, now succeeded to the command.   General Wade Hampton,
second in command, a South Carolinian, refused to
serve under " the scamp."   Wilkinson flourished a
few proclamations of which this general order is a
sample : " The plan of campaign is to rendezvous the
whole troops on the Lake in the vicinity of Sacketts
Harbor, and in cooperation with our squadron, to make
a bold feint upon Kingston, slip down the St. Lawrence, lock up the enemy in our rear to starve or surrender ; sweep the St. Lawrence of armed craft, and in
concert to take Montreal."   This was all that he proposed to do at once.   Eventually he dawdled and
quarrelled away the summer, and after a slight advance
with a strong force, he was whipped by 200 of the
enemy.   He was then displaced by the Secretary of
War ; and afterward arrested and court-martialed.
The country was finally rid of the wretched scoundrel.
Twice court-martialed, a pensioner of Spain, the ally
of Burr, a coward, an incompetent, only a strange combination of circumstances had kept him from being shot
long before his retirement.   At last rid of incompetents
and cowards, the campaign of 1814 was fought on very
different terms.   The battles of Lundy's Lane and
Chippewa gave real glory to our soldiers ; and General
Winfield Scott came to be a name honored in every
American household.   But at the close of the year each

party remained on its own soil. Nothing had been gained, or lost.

Along the ocean coast the British had adopted a war of incursions, plunder, and torch. They had burned not only several towns in Connecticut and elsewhere ; but had occupied Washington, driven out the government, and burned the capital. As a final stroke England had planned to concentrate a large force and invade by way of Louisiana. It was believed that there was still much disaffection toward the United States in that quarter.

Throughout the Union, in spite of disasters and blunders and hardships, loyalty grew in force. In New York the party of the administration was stronger than at the opening of the war. Kentucky, which had been so singularly loyal during Burr's plot, begged to be allowed to send troops to aid at Detroit. Col. Johnson of that State was the conspicuous fighter when Harrison led the way into Canada. Everywhere a determined front of the people was seen, except in the East. No son of New England can remember without pain and shame the record of that section. " In fact my Lord," wrote Provost to Bathurst, " two-thirds of the army in Canada are at this moment eating beef provided by American contractors." Says MacMaster, " The road to St. Regis was covered with droves of cattle, and the river with rafts destined for the enemy. On the Vermont side of the lake the highways were too narrow and too few to accommodate the herds of cattle that were pouring into Canada." " Were it not for these supplies," wrote General Isard to the Secretary of War, " the British forces would soon be suffering famine." England in return exempted Massachusetts, Connecticut, and New Hampshire from blockade.

The plan was now discussed by the Federalist leaders of withdrawing all New England troops to their own soil. Governor Chittenden of Vermont issued orders to Vermont's regiments in New York to return home. The officers read the proclamation to the troops, and they all united in sending back word to Chittenden that they would not obey him. "We are," said they, "in the service of the United States, and your power over us as governor of Vermont is suspended. We will continue in the service of our country till discharged. We regard your proclamation with mingled emotions of pity and contempt." The fact was the soldiers were mostly Republicans. Notwithstanding the efforts of the Federalists, Massachusetts had furnished a generous quota of troops during the whole war ; but Governor Strong in a message declared our government the offender, as against Great Britain. After Lawrence's splendid sea fight the whole nation held a holiday ; but the Massachusetts legislature passed a resolution that "it did not become a religious people to express any approbation of military or naval exploits not immediately defensive." This was twaddle. It showed however that the peace party did not dare overt treason. The war party was too strong. Commodore Decatur detected blue light signals to give the enemy warning of his movements in New London harbor. From that time "Blue-Light Federalist" became a party sobriquet and stigma.

The irrepressible but detestable Timothy Pickering bobbed once more into publicity. He presided at a meeting in Essex Co., which passed resolutions, drawn up by himself in favor of a State convention. This proposition, looking toward a revival of the old disunion project, was supported elsewhere, and notably by Otis ;

but Dexter who had formerly been of the same spirit
fought it now to the death. Josiah Quincy declared in
Congress that the President's " Cabinet, little less than
despotic, was composed of two Virginians and one
foreigner." (Madison, Monroe, and Gallatin.) Finances became embarrassed to the point of collapse.
The banks, except those of New England, which had
refused to loan to the government, came to an inevitable crash. Government treasury notes were worth
seventy cents on a dollar. The peace party was gaining strength in New England. Young Daniel Webster's first appearance in Congress was as a bitter antagonist of the war. At Rockingham in New Hampshire
he had offered the following equivocal Resolution :
" We shrink from the separation of the States as an
event fraught with incalculable evils. If a separation
ever should take place it will be on some occasion
when one portion of the country undertakes to control,
to regulate, and to sacrifice the interests of another.
The government may be assured that the tie which binds
us to the Union will never be broken by us." In Congress he appeared as an active opponent of the war and
the administration. New England was carrying on
all this while a coast trade of considerable size with the
British troops, and with the West Indies ; while Southern ports were blockaded by the British fleets. Congress at last in its indignation laid an embargo against
all articles known to come from Great Britain. This
shut up every port in a trice, and stopping their gains,
added immensely to the anger of New England. The
wisdom of embargoes is now pretty well known ; and it
is not likely they will ever again be resorted to in our
history. In the present case only fuel was added to the
flame. The Federalist leaders began to have the

people of the Eastern States almost unanimously at
their backs. The embarrassed administration had
made many blunders ; and the embargo was one of
them. The old dethroned leaders were once more apparently monarchs. Otis introduced a resolution in the
Massachusetts legislature offering aid to Chittenden of
Vermont in enforcing the laws against the soldiers of
that State. The proposition was discussed of forming
a separate treaty with Great Britain.

The value of the militia of New England was largely
negatived by the well known sentiment of her leaders,
that as little aid as possible should be rendered outside
their own State. They were enlisted mostly only on
paper ; but on that paper they afterward drew thousands
of dollars pension from the national treasury. Nothing
has ever so debauched and pauperized a people as a
general system of pensions. The Governor of Connecticut in August withdrew all the State militia from the
command of national officers, and ordered them to obey
only a major-general of militia. On that very day,
August 24, Washington was captured by the British.
Five days later the banks of Philadelphia suspended
payments.

New England was practically in rebellion. It had
seceded from united national action, and had set up a
war confederacy. Whether it could compact a peace
organization remained to be seen. Governor Strong
in October called together the legislature, and said to
them that the national government had failed to fulfil
the terms of the Constitution, to protect Massachusetts
from invasion or attack. They must henceforth look
to God and themselves. He more than hinted that the
time had come for a separate New England alliance.
The Boston Centinel declared the Union was as good

as dissolved. Pickering chirped from Washington,''
'' Abandoned by the general government, except for
taxation, we must defend ourselves ; so we must secure
and hold fast the revenues.'' The legislature promptly
responded to Governor Strong of Massachusetts that
there remained no alternative before them but either
to submit to the foreign enemy or control their own
resources. It farther said that the Constitution had
failed to be of the benefit to New England that had been
expected ; and that it must be supplanted.

Madison was certainly not to be envied. He had
inherited this war, which had been staved off by his
predecessors only to fall on his administration. William Wirt writes that he called upon the President,
October 14th. He found him in a temporary residence.
'' He looks,'' wrote Wirt, '' miserably shattered and
woebegone. In short he looks heart broken. His
mind is full of the New England sedition. He introduced the subject and continued to press it, painful as
it obviously was to him.'' Jefferson wrote that the de-
fection of Massachusetts was not fatal. '' If they become neutral we are sufficient for one enemy without
them; and in fact we get no aid from them now.'' It
is a curious fact that about this time a proposition was
discussed in New England to form an alliance with
South Carolina to resist Virginia. So strong was the
similarity of the two sections in temper, religion, and
trading instincts. Calhoun of all Republican statesmen was most emphatically opposed to the continuance
of the embargo. Think of the possibility of such an
alliance!

Not however till October of 1814 were measures of a
positively treasonable sort inaugurated. The very day
that dispatches from Ghent announced the peace pro-

posals of England, the Massachusetts legislature issued
an invitation for a Conference of New England States
to be held at Hartford.   Connecticut and Rhode Island
promptly responded.   The Boston Centinel spoke of
Massachusetts, Rhode Island, and Connecticut as the
first three pillars '' in a new Federal edifice.''   It was
proposed to make a special and separate treaty with
England.   In the Conference at Hartford there were
twenty-six delegates, representing not only the three
States named, but parts of Vermont and New Hampshire.   Precipitate action was checked by the character
of two or three of the leaders.   As they came close to
overt secession they quailed before possible consequences.   The Convention discreetly sat with locked doors.
New England was far from being a unit behind them.
It was not to be expected that secession would be peaceably acquiesced in by other States.   Newspapers were
largely in favor of prompt action.   Pickering was spur-
ring the delegates.   Gouverneur Morris, a fussy creature who had more influence than he deserved, wrote,
'' If not too tame and timid, you will be hailed hereafter
as the patriots and sages of your generation.''   Popular
gatherings resolved to pay no more national taxes until
the Convention had decided the course of New England.   Pickering wrote, '' If the British succeed in their
expedition against New Orleans ; and if they have
tolerable leaders I see no doubt of their success ; I
shall consider the Union as severed.   I do not expect
to see a single representative in the next Congress from
a Western State.''   He was doomed to disappointment.
He is the only member of an American cabinet who
ever died in his bed that deserved to be hanged.   In
Congress he was measured at his real worth.   Governor Wright of Maryland resolved on impeaching him

for treason. With a pile of books under each arm he
met Randolph, who asked what it meant. "It means,"
said the governor, choking with rage, "It means—
Timothy Pickering! I will convict him of treason."
"But sir," said Randolph, "you do not mean to attack
him without due notice." The governor, who was the
ideal of chivalry, said, "Sir! do you think etiquette
demands it?" Stalking down the aisle with his arma-
ment, he nudged Pickering with his elbow—then pointing with his thumb to an armful, he said, "Do you see
that?" "Yes sir!" Then pointing to the other
arm-full, with the other thumb, he said, "Do you see
that?" "Yes sir, but what does it mean?" "It
means," said the governor with great formality, "that
with these I mean to give you what you richly deserve. . . ."

In the Hartford Convention Cabot and Prescott advised caution. They saw that to merely cut loose from
the Union was not to create a new nation. They
doubted the consequences of precipitate action. To a
young delegate Cabot, who was made presiding officer,
said, "We are going to keep you young hot-heads
from getting into mischief." Vermont was loyal and
Republican. Such men as General Varnum of Massachusetts in Congress were true to the existing government. The members received letters of warning; and
Lowell says, "Otis received every day or two notes
threatening him with bodily harm." "A thousand
fears restrain him." Any overt act of treason would
probably have caused instant resistance. Civil war
would have broken out that they were unprepared to
quell. The large majority against the war, did not
mean that as large a majority would sustain disunion.
The two parties grew so embittered that, at town

meetings, held in town churches, they would gather
on opposite sides of the broad aisles, consulting together, but having nothing to do with their opponents.

Secresy, while working well to give freedom of debate
in the Convention and to conceal private opinions, was
not working well outside. The people of New England
were suspicious of conspiracy. They talked a great
deal on surmises ; and this public discussion was hot
and testy. Every one was at liberty to charge on the
convention the most extreme and outrageous proceedings. No one could authoritatively reply. The loyalists had the advantage, and they saw it and used it.

The Convention reported among other things that
" If the Union be destined to dissolution, some new
form of confederacy should be substituted among those
States which shall intend to maintain a federal relation
to each other." " A severance of the Union by one
or more States especially in time of war can be justified
only by absolute necessity." " That Acts of Congress
in violation of the Constitution are absolutely void is
undeniable." This nullification principle however it
would modify by " It does not however consist with
respect and forbearance, due from a Confederate State
toward the general government, to fly to open resistance upon every infraction of the Constitution."
Amendments to the Constitution, seven in number,
were proposed ; and three Resolutions of demand on
Congress. If the demands were not granted them,
" Resolved that it will be expedient for the legislatures of the several States to appoint delegates to
another convention to meet at Boston, with such
powers and instructions as the exigency of a crisis so
momentous may require." What this meant it is not
needful to suggest. It was nullification, followed by a

threat of secession and coercion. " State action " was
proposed to protect New England against drafts and
conscriptions for soldiers to fight against England. An
arrangement was demanded whereby a portion of the
taxes collected for the general government should be at
the disposal of the States, for such individual conduct
of the war as each might choose. It was also proposed
that New England create State armies for self-defense.

The Federalists in Congress were incessant in urging
extreme action. " How," said Pickering, " are the
powers reserved to the States respectively, or to the
people, to be maintained, but by the respective States
judging for themselves, and putting their negative on
the usurpations of the general government ? " This
sort of talk was not confined to this country. It was
published in England.

In Massachusetts the legislature, in session at Boston, passed a resolution to send a delegation to demand the taxes as proposed by the convention. Harrison Gray Otis was the head of this committee.
Governor Strong who had been all along bitterly
Federal, undertook at once to raise a State army, and
in every way to create an independent commonwealth.
But the utmost amount of money he could secure was
631,000 dollars. It was proposed to seize the national
taxes. At the same time the State refused to cooperate
with the government in driving the British from Maine.
John Adams, over eighty, bristled like a porcupine at
this state of affairs. He declared Cabot, Otis, Strong,
and others were only moved by selfish ambition. It
was " a plot of rogues " to rule. Adams had parted
from the Federalists to accept of democracy. Jefferson
in turn had parted from extreme democracy to accept
of a degree of centralized government that brought him

very near to Adams. Both had been to school to experience. It is not wonderful that they in old age forgot early alienations, and lived, as they died, in unison.

Connecticut followed on the heels of Massachusetts. What all this would have amounted to within six months needs no prescience to comprehend ; only that New England was sorely put to shame by two events. The British who marched on New Orleans got into the very suburbs of the city ; while Jackson had been fooled by them to the last minute. But while he was too hot and stubborn to be a good field marshal, he was a terrible fighter. He reached New Orleans in time. He swept hindrances out of the way with a high hand ; and then nearly annihilated the British. The victory was complete. Meanwhile at Ghent had been signed the most extraordinary treaty England was ever cajoled to execute. Beginning with high and lofty demands, her commissioners had been crowded by the Americans to yield at one point and another, until we had won a triumph of diplomacy far ahead of our triumph at arms. Our commissioners John Quincy Adams, James A. Bayard, Henry Clay, and John Russel, were headed by that most consummate statesman Gallatin. The British Commissioners were men of hardly average ability, and were flanked at every move. The points first demanded by England were that America must yield all the Northwest, including Michigan, Wisconsin, Illinois, a large part of Indiana, and one-third of Ohio as a perpetual Indian Territory—a barrier between Canada and the United States ; that we must also renounce our right to keep armed vessels on the lakes, or military ports on the shores ; and thirdly we must relinquish a considerable portion of Maine to be

15

British property. It was these terms that the members of the Hartford Convention, and other New Englanders, had declared to be "just and liberal." But the rest of the States indignantly had spurned the proposals, and voted to increase armies and supplies. Adams wrote to Madison to "continue the war forever rather than yield one acre of territory, or the fisheries, or impressment." Our commissioners talked firmly and unyieldingly. Little by little the English gave way on all points except the Maine territory. The English cabinet proposed to Wellington to go to America as commander-in-chief, with power to make peace, or fight. He bluntly told them to make peace themselves; for they had done nothing to warrant them in demanding territory of the States. Nothing at last was left but a claim on the part of England for a free use of the Mississippi, and an equivalent for fishing off Newfoundland. But these in turn were yielded; for the English were more anxious for peace than they dared avow.

Nothing more ludicrous in American history ever rippled along the surface than the good natured fun made of the New England delegates who were in Washington demanding practically separation from the Union. All that they had presumed and built on had not occurred. Not one of their premises proved correct. The English people had backed down completely. Peace had been declared; but not a single British demand had been yielded — notwithstanding the votes at New England elections and the resolutions of conventions that pronounced the demands just; and to grant them to be political wisdom. New Orleans had not been taken. The delegates who could in the face of such events obey instructions would only have increased public ridicule. They went home as quietly as possible. The rebels who had so nearly led their States into treasonable conduct were never again heard from ; and New England from that day became among the faithful most faithful to the Union.

But it was now, after the hated war was closed, that the Eastern States began to suffer their worst trial. Public sentiment was tinged with a spirit of indignation, and there was some inclination to retaliate on the section that had made so much trouble in the hour of most need. The Secretary of the Treasury crowded New England in the collections of national reserves ; but was more lenient to other sections. England poured in the contents of an enormous glut of manufactures, and undersold even the highly protected home manufactures. Shipping was soon compelled to compete with a very great increase of foreign vessels. There was an era of depression and general suffering. The people however gave little sympathy. The word Yankee became a byword. But New England had one relief ; she could emigrate. There is a hand of Providence in every period of history, shaping events to ultimate betterment. Certainly no event in American development and nation-building equals in importance that outflow of New England men and institutions to take possession of the great Northwest. Already Western New York, beyond Utica, was a second Connecticut. Massachusetts had largely settled southern Ohio ; and Connecticut northern Ohio. The Reserve, from Erie to Cleveland, was known as " New Connecticut." Michigan was also a child of the six New England States. As the tide moved westward and severer colors of New England life faded out, there was less bigotry and more toleration. But most important was it that the common school and the town meeting were the basis of every colony. These were after all the soul of New England ; the noblest inheritance of her sons and the dower of her daughters. There had never been a fondness for making trouble shown by the industrious tactful common folk of the Eastern States. The leaders were left behind ; their power breaking ; their supremacy in State and Church about to pass away forever. But in spite of migration New England suffered wofully. It had a local financial crisis by itself ; and it did not get the sympathy that would naturally have been its share. Adams says, " So discredited was Massachusetts that she scarcely ventured to complain ; for every complaint by her press was answered by the ironical advice that she should call another Hartford Convention."

The war had been fought under terrible disadvantages ; but it had closed as having accomplished more permanent advantage to civilization than any other two years of history. It made of the oceans a vast republic ; and established " the freedom of the seas." It established the rule of " free ships free goods ; " and brought the right of search under the restraint of just international law. It established the great doctrine of individual rights ; and forever ended the tyrannical code of " once a citizen always a citizen." Henceforth a man who felt that he could better his condition could transfer his residence to another land and therewith his fealty. The United States alone, and against great odds, had made the most important enlargement of the Code of Nations that had ever been recognized as international law.

It is impossible to estimate the value of the treasonable episode of 1814 in lowering the pride and breaking the authority of " the Best." It was equally useful in loosening the hold of bigotry. Creeds began to be amenable to reason ; and blue laws lay in the Statute books unobservable. It is a curious fact that from about 1815, immediately after the war, New England's whole intellectual power turned into bold free investigation. Parker and Channing and Emerson and Garrison, the Transcendentalists and Unitarians throve as nowhere else in the States.

Unitarianism as well as democracy grew aggressive. It was popular freedom in Church and in State. Channing, at first inclined to resent the idea of being an innovator, found that after all he was a lamb leading lions. The people ran ahead. The clergy no longer could hold the people under a conceived obligation to be damned for the glory of a New England God. Civil power was rapidly divorced from church power ; and the natural historic collateralism of Church and State was resumed.

The course of the administration had throughout the war been moderate and patient. Jefferson wrote to General Dearborn, " Oh Massachusetts ! how have I lamented the degradation of your apostasy ! Massachusetts with whom I went with pride in 1776 ; whose vote was my vote on every public question, and whose principles were then the standard of whatever was free or fearless. But then she was under the counsels of the two Adams ; while Strong her present leader was promoting petitions for submission to British power and usurpation. Let us look forward to her dismissing venal traitors as the signal of return to the principles of her brethren : and if her humiliation can just give her modesty enough to suppose that her Southern brethren are somewhat on a par with her in wisdom, in information, in patriotism, in bravery, and even in honesty, she will more justly estimate her relative mo-

mentum in the Union." He did not believe Massachusetts had intended final separation. " The majority
of the leaders did not aim at separation. In this they
adhered to the known principle of General Hamilton,
never under any views to break the Union. Anglomany, monarchy, and separation were the principles
of the Essex Federalists ; Anglomany and monarchy
were those of the Hamiltonians ; and Anglomany only,
that of the people who called themselves Federalists.
These last were as good Republicans as the brethren
they opposed ; and differed from them only in devotion to England and hatred for France. The moment
the leaders avowed separation from the Union the people quit them to a man." Felix Grundy called the
action of the Federalists " moral treason " ; an expres-
sive description, which Calhoun adopted and endorsed.

The war was not the real cause of the disaffection.
Lowell wrote in the Centinel : " There are men who
know that our troubles are not the offspring of this war
alone ; and will not die with it. But they despair of
relief. They consider the people, in their very nature
democratic—that should they be restored to prosperity, the same passions and opinions would diffuse
themselves through the country." Cabot in disgust
wrote to Pickering, " Why can't you and I let the
world ruin itself in its own way ? " There was not in
their minds the slightest doubt but the world would
ruin itself but for them ; wisdom would die with them.
Democracy was " disease," and it had become a New
England disease. The only salvation was to devise
some method of getting out of the Union ; and then set
up as soon as possible a rule of " the Best." No
theory to defend secession was undertaken, because so
far none was necessary. The effort to break away and

create a Northern Confederacy was secretly placed on
the fact that in no other way could democracy be suppressed. In public the reason given was that the
Union did not foster their prosperity—that it did not
protect them, while calling on them to protect others.

A just summary of the situation shows the Federal
leaders struggling to distract the United States in the
conduct of the war ; to disable the general government
both as to money and troops ; to place us at a disadvantage with England, until forced to make peace on
such terms as the enemy proposed. There was a total
lack of patriotism ; and a vigorous pushing of provincialism ; a refusal to contribute financial aid, but a
demand for such aid for local defense. Worst of all
was the badly concealed rejoicing over the straits of
the nation. Practically New England had been in rebellion throughout the whole war.

The guilt lay with the old leaders who had already
twice proved their unfaithfulness to the Union. These
men, moved quite as much by theological rancor as by
political prejudice, had concentrated all their hate and
abhorrence on Thomas Jefferson. The war was " unquestionably his work. Madison was a man not bred
to camps or naturally prone to contests." The comic
feature of the controversy was the determination of the
leaders to see or at least assert that they could see, an
alliance of Thomas Jefferson and Napoleon Bonaparte.
It was true that Jefferson was no longer in office ; but
was not Madison under his influence ?

With this wretched display of treachery Federalism
vanished forever from American politics. There had
been little chance for loyalty to the whole United
States on the part of New England so long as the old
leaders remained alive. Fortunately they were now

rapidly dropping off. The younger men like Webster
had soon to face threats of nullification and secession in
other quarters ; and to put themselves in training as
defenders of the Constitution. Pickering spent the
remnant of his life in maligning John Adams, writing
notes refusing to meet at table President John Quincy
Adams, and sneering at the Declaration of Independence. With a heredity of several generations of self-
glorification in being " chosen of God," he mixed religion and politics with such skill that he could baptize
any meanness, and serve the Lord with any dishonor.
He openly expressed contempt for Washington and
hatred for Adams, while Jefferson his whole saintly
soul abominated. He was unquestionably a pious
man ; never omitting his prayers, and rarely his objur-
gations. Lodge believes that Pickering did not desire
separation except as a means " to coerce the other
States into submission to New England principles."
Governor Strong consoled himself with the belief that
even though they had failed ; and the war had ended
gloriously, it was not to be supposed the Western
States would long continue in the Union. " They
will," he said, " soon prefer a government of their
own. We may be happy as neighbors, when a Union
would be inconvenient."

The right of a State or section to withdraw from the
Union was even yet not disputed. At all events Massachusetts and Connecticut, with their militia scarcely
diminished by the war, could have defied the exhausted
forces of the rest of the States. There is little probability that coercion would have been attempted to retain them in the Union. New York and New Jersey as
border States had been faithful to the government ;
but it is not improbable they would have refused to

allow national troops to cross over their soil for hostile purposes into that of their Northern neighbors. Western New York was filled with Connecticut pioneers and northern New York was an overflow of Vermont. Loyal to the core, yet they loved the parent States.

The treaty of Ghent was a victory as sweeping as that of Jackson at New Orleans. It began the era of a better understanding between England and America. The dogged determination of Great Britain to refuse to us the rights of a first class power waned. The struggle in Europe to sustain commercial war in time of peace had been tested to its utmost. Napoleon had crowded the policy of national embargoes and laws against free trade to outrageous extremes. His effort to crush the commerce of England had been his own overthrow. Civilization began to doubt the wisdom of imposts between nations. The States had established free trade for this continent. Saxon blood was warming. Castlereagh who had assented to the despotic movements of the Bourbons, committed suicide. The great Canning had succeeded him as prime minister. A new era of good will was about to dawn upon America and Great Britain.

APPENDIX TO CHAPTER V

REPORT OF THE HARTFORD CONVENTION, DATED JAN. 4, 1815

. . . It is a truth, not to be concealed, that a sentiment prevails to no inconsiderable extent, that administration hath given such constructions to that instrument (Constitution), and practiced so many abuses under color of its authority, that the time for a change is at hand. Those who so believe, regard the evils which surround them as intrinsic and incurable defects in the Constitution. They yield to a persuasion, that no change, at any time, or on any occasion, can aggravate the misery of their country. This opinion may ultimately prove to be correct. But as the evidence on which it rests is not yet conclusive, and as measures adopted upon the assumption of its certainty might be irrevocable, some general considerations are submitted, in the hope of reconciling all to a course of moderation and firmness, which may save them from the regret incident to sudden decisions, probably avert the evil, or at least insure consolation and success in the last resort. . . .

Finally, if the Union be destined to dissolution, by reason of the multiplied abuses of bad administrations, it should, if possible, be the work of peaceable times, and deliberate consent. Some new form of confederacy should be substituted among those states which shall intend to maintain a federal relation to each other. Events may prove that the causes of our calamities are deep and permanent. They may be found to proceed, not merely from the blindness of prejudice, pride of opinion, violence of party spirit, or the confusion of the times ; but they may be traced to implacable combinations of individuals, or of states, to monopolize power and office, and to trample without remorse upon the rights and interests of commercial sections of the Union. Whenever it shall appear that these causes are radical and permanent, a separation, by equitable arrangement, will be preferable to an alliance by constraint, among nominal friends, but real enemies, inflamed by mutual hatred and jealousy, and inviting, by intestine divisions, contempt and aggression from abroad. But a severance of the Union by one or more States, against the will of the rest, and especially in a time of war can be justified only by absolute necessity. These are among the principal objections against precipitate measures tending to disunite the States ; and when examined in connection with the farewell address of the Father of his Country, they must, it is believed, be deemed conclusive. . . . That acts of Congress in violation of the Constitution are absolutely void, is an undeniable position. It does not, however, consist with the respect and forbearance due from a confederate state towards the general government, to fly to open resistance upon every infraction of the constitution. The mode and the energy of the opposition should always conform to the nature of the violation, the intention of its authors, the extent of the injury inflicted, the determination manifested to persist in it, and the danger of delay. But in cases of deliberate, dangerous, and palpable infractions of the Constitution, affecting the sovereignty of a state, and liberties of the people ; it is not only the right but the duty of such a state to interpose its authority for their protection, in the manner best calculated to secure that end. When emer-

gencies occur which are either beyond the reach of the
judicial tribunals, or too pressing to admit of the delay
incident to their forms, states which have no common
umpire, must be their own judges, and execute their
own decisions. . . .

RESOLUTIONS OF THE HARTFORD CONVENTION

Therefore resolved, that it be, and is hereby recommended to the legislatures of the several states represented in this Convention, to adopt all such measures
as may be necessary, effectually to protect the citizens
of said states from the operation and effects of all acts
which have been or may be passed by the Congress of
the United States ; which shall contain provisions, subjecting the militia or other citizens to forcible drafts,
conscriptions, or impressments, not authorized by the
constitution of the United States.

Resolved, That it be, and hereby is recommended to
the said Legislatures, to authorize an immediate and
earnest application to be made to the government of the
United States, requesting their consent to some arrangement, whereby the said states may, separately or in
concert, be empowered to assume upon themselves the
defence of their territory against the enemy ; and a
reasonable portion of the taxes, collected within said
States, may be paid into the respective treasuries
thereof, and appropriated to the payment of the balance due said states, and to the future defence of the
same. The amount so paid into the said treasuries, to
be credited ; and the disbursements made as aforesaid,
to be charged to the United States.

Resolved, That it be, and hereby is, recommended to
the legislatures of the aforesaid states, to pass laws

(where it has not already been done) authorizing the governors or commanders-in-chief of their militia to make detachments from the same, or to form voluntary corps, as shall be most convenient and conformable to their constitutions, and to cause the same to be well armed, equipped, and disciplined, and held in readiness for service ; and upon the request of the governor of either of the other states, to employ the whole of such detachment or corps, as well as the regular forces of the state, or such part thereof as may be required, and can be spared consistently with the safety of the state, in assisting the state, making such request to repel any invasion thereof which shall be made or attempted by the public enemy.

Resolved, That the following amendments of the Constitution of the United States be recommended to the states represented as aforesaid, to be proposed by them for adoption by the state legislatures, and in such cases as may be deemed expedient by a convention chosen by the people of each state.

And it is further recommended, that the said states shall persevere in their efforts to obtain such amendments, until the same shall be effected.

First. Representatives and direct taxes shall be apportioned among the several states which may be included within this Union, according to their respective numbers of free persons, including those bound to serve for a number of years, and excluding Indians not taxed, and all other persons.

Second. No new state shall be admitted into the Union by Congress, in virtue of the power granted by the Constitution, without the concurrence of two-thirds of both houses.

Third. Congress shall not have power to lay any

embargo on the ships or vessels of the citizens of the United States, in the ports or harbors thereof, for more than sixty days.

Fourth.   Congress shall not have power, without the concurrence of two-thirds of both houses, to interdict the commercial intercourse between the United States and any foreign nation, or the dependencies thereof.

Fifth.   Congress shall not make or declare war, or authorize acts of hostility against any foreign nation, without the concurrence of two-thirds of both houses ; except such acts of hostility be in defence of the territories of the United States when actually invaded.

Sixth.   No person who shall hereafter be naturalized, shall be eligible as a member of the senate or house of representatives of the United States, nor capable of holding any civil office under the authority of the United States.

Seventh.   The same person shall not be elected president of the United States a second time; nor shall the president be elected from the same state two terms in succession.

Resolved, That if the application of these states to the government of the United States, recommended in a foregoing resolution, should be unsuccessful, and peace should not be concluded, and the defence of these states should be neglected, as it has been since the commencement of the war, it will, in the opinion of this convention, be expedient for the legislatures of the several states to appoint delegates to another convention, to meet at Boston in the state of Massachusetts, on the third Thursday of June next, with such powers and instructions, as the exigency of a crisis so momentous may require.

Resolved, That the Hon. George Cabot, the Hon,

Chauncey Goodrich, and the Hon. Daniel Lyman, or any two of them, be authorized to call another meeting of this convention, to be holden in Boston, at any time before new delegates shall be chosen, as recommended in the above resolution, if in their judgment the situation of the country shall urgently require it.

ACTION OF TOWNS

Amherst adopted the following Resolution Jan. 3, 1814, Noah Webster presiding :

That the representatives of this town in the General Court are desired to use their influence to induce that honorable body to take the most vigorous and decisive measures compatible with the Constitution to put an end to this hopeless war, and to restore to us the blessings of peace. What measures it will be proper to take, we pretend not to prescribe ; but whatever measures they shall think it expedient to adopt, either separately or in conjunction with the neighboring States, they may rely upon our faithful support.

Newbury, a town in Pickering's district, voted that : We remember the resistance of our fathers to oppressions which dwindle into insignificance when compared with those which we are called on to endure. The rights " which we have received from God we will never yield to man." We call on our State legislature to protect us in the enjoyment of those privileges to assert which our fathers died, and to defend which we profess ourselves ready to resist unto blood. We pray your honorable body to adopt measures immediately to secure to us especially our undoubted right of trade within our State. We are ourselves ready to aid you in securing it to us to the utmost of our power, " peaceably if we can, forcibly if we must " ; and we pledge to you the sacrifice of our lives and property in support of whatever measures the dignities and liberties of this free, sovereign, and independent State may seem to your wisdom to demand.

It has been suggested that Pickering was himself the author of several of these fulminations.

CHAPTER VI

SOUTH CAROLINA NULLIFICATION IN 1832

THE era of factions as we have seen began as soon as the Union was formed. It continued through the administrations of Washington, Adams, Jefferson, and Madison. The English party and the French party, with constantly shifting degrees of zeal, divided the people. Other questions complicated matters, but at the bottom it was, down to the war of 1812–14, a factious antagonism between those who would submit to English imposition and those who would not. New England had been the stronghold of the English faction not only because it was New England, but because it had so far been, not a manufacturing, but a trading section, and its trade was in the main with Great Britain. The second war with England had ended in such a manner as to greatly foster national unity, and to lessen sympathy for the power we had twice crippled. Unification and homogeneity of sentiment had been vastly increased by the wise and generous administration of Jefferson. Monroe's administration, extending from 1816 to 1824, was the era of calm and peace. Federalism, dead as a party, was now also dead as a factional issue. The Spanish faction was assimilated. The English faction had faded out.

There was at last an American people rapidly becoming autonomous. The colonies had become cemented into a nation. The process of opening great turnpikes and canals was going on with rapidity, to connect the different sections. Calhoun and Webster and Clay were soon to be the leaders at Washington, in place of Jefferson, Madison, and Monroe. They were not Federalists ; neither were they Republicans.

During all the era of factionalism there had been more or less also of sectionalism. It had at times shown a vicious readiness for making disturbance. It entered strongly into the attempts of New England to secede. Still it had been possible to elect peaceably four out of the first five Presidents from Virginia. Monroe, the last of these, was as popular in New England as in the Carolinas. Jefferson had met Adams half way, and their last days were of one mind,—till their spirits flew away together as a symbol of national harmony. The fact was that through all this period the Federalists were for the most part right-hearted while they were wrong-headed. We were sons of England. Blood is thicker than water. We never intended to give up our Anglo-Saxon heritage, or our English language. To pull the root of our institutions out of English life and history we could not if we would. Our instincts, our prejudices, our vices, and our virtues were of the English stock. The only trouble was that England had not yet learned to meet her children as part and parcel of herself. The paternalism that made her hold the rod over us must be given up. The Federalists had been inclined to submit to the rod ; the Republicans refused.

In 1823 occurred the final act in this era of faction. It was the more welcome that it was unexpected.

Through the Napoleonic wars with England we had suffered outrage and spoliation from both powers, because of our neutrality. But Napoleon was at last dead in his cage at St. Helena, and the European despots were restored to their thrones. A coalition of Russia, Austria, and Prussia, called the Holy Alliance, was formed to sweep out of Europe all remnants of popular government. Dei gratia was once more the watchword ; the people were to be ruled in the name of God, both in state and church. Louis XVIII. of France although not of the Alliance was of the same temper. While he suppressed popular government in Naples he marched his armies into Spain and restored the Bourbons. A conference was summoned to meet at Paris to consider next the condition of affairs in America. The South-American dependencies of Spain had revolted and set up republics. The despots proposed to reduce these, and divide them among themselves. This could not have been done without fatal consequences to the United States. The battle was being drawn between popular government and legitimacy. It took in all the world. The odds were terribly against republicanism and freedom. Castlereagh had committed England to the legitimists. But Castlereagh committed suicide, and Canning was called to the ministry.

Now occurred what was truly the greatest event of the nineteenth century. Canning refused to send delegates to Paris ; and instead turned to the United States and said, "The hour has come when once more we must stand together." To the legitimists he said : "Your opinion that all power is derived from God-appointed monarchs strikes at the fundamental principle of the British constitution. It is false. We will not deny that all power originates with the people."

In other words Canning withdrew England from the alliance with despotic governments, and proposed a limited alliance with the United States. The English-speaking race in his judgment was once more called to fellowship and to co-operate in the name of liberty. The whole history of England had been violently wrenched to bring us into conflict. Let us once more become a united folk.

Our minister, Mr. Rush, hastened dispatches to Washington, containing Mr. Canning's proposition. Monroe and his cabinet were astounded beyond measure. John Q. Adams, Secretary of State, rose to a full measure of the greatness of the occasion. Calhoun endorsed it with the fiery enthusiasm of his youth. But Jefferson was still alive, and was reverenced as the Sage of Monticello. The documents were forwarded to him for his advice. In October of 1823 he answered as follows : '' The question presented by the letters you have sent me is the most momentous which has been offered my contemplation since that of Independence. That made us a nation ; this sets our compass. While Europe is laboring to become the domicile of despotism our endeavor should be to make our hemisphere that of freedom. One nation most of all can disturb us in this pursuit. She now offers to lead, aid, and accompany us. By acceding to her proposition we detach her from the band of despots, and bring her mighty weight into the scale of free government. With Great Britain withdrawn from the scale, and shifted into that of our two continents, all Europe combined would not undertake war.'' It was a glad hour for this great mind.

The advice of Jefferson was followed. Mr. Monroe, as Canning requested, took the initiative, and in De-

cember sent a message to Congress in which he said,
" We should consider any attempt on the part of the
Holy Alliance to extend their system to any nation of
this hemisphere as dangerous to our peace and safety."
This was the Canning-Monroe Doctrine, or as Jefferson
called it, " the American System." It was the proclamation that America and England formed once
more one people, with one great mission, and that together we should civilize the earth.   The era of faction
ended in America.   The greatest marvel is that this
sublime event, hardly surpassed in nobility of purpose
and consequences by any event in the history of man,
has been so little apprehended by the American people.
The era of sectionalism was already upon us ; and politicians observing the surface facts only, saw in the Monroe Doctrine defiance hurled by America at all the
world.   It has even been assumed of late to involve a
protectorate of the American continents.   Meanwhile
the Canning proposition enabled the United States to
move forward, fearless of all foreign interference, to
spread a homogeneous population from ocean to ocean ;
while England carried a civilization equally great into
Asia, Africa, and Australia.   It was this alliance that
led on to the fact that English-speaking Anglo-Saxon
ideas now gird the globe ; while the language of Mil-
ton is the language of one-third of the human race ;
and the hope of a Federal Union, linking all continents, is not as far away as, in 1823, was the union of
fifty States, bound by steel rails and English instincts.

We pass, by the way of this transitional event, into
the second era of American history.   Two forces were
already at work long before the end of the era of factions to create the era of sectionalism.   Tariff legislation and slavery were twins in bringing into antagonism

the North and the South. The Louisiana purchase and
the invention of the cotton gin created a great slave
market which Jefferson in vain tried to close. Virginia
found it more profitable to breed slaves than to grow
tobacco. Instead of being abolitionists like Jefferson
and Madison and Monroe, her statesmen became first
apologists for slavery, and then its advocates. South
Carolina, which never separated religion and politics,
became economically and piously attached to the
domestic customs which were now to be excluded from
all territory north of Virginia and Kentucky. New
York closed up the Northern platoon of freedom by
abolishing slavery in 1817 ; the act to be completed in
1827.

Slavery began to take possession of legislation at the
South. Virginia in 1819 threatened the enslavement
of all free blacks. Georgia taxed every free negro
twenty dollars a year, and expelled him from the State
for non-payment of taxes. This was a dangerous edge
to an endless wedge—for injustice knows no limit to
its demands. Mississippi and Alabama forbade legislative emancipation. If the free negro fled across the
Ohio he was arrested as a tramp and imprisoned.
There was no chance for him. A dozen States forbade
the teaching of a black man the alphabet. The slave
trade, notwithstanding the act of 1807 prohibiting it,
was still carried on. In 1819 a more stringent act was
passed. Even Northern negroes were kidnapped and
sold at the South. Jefferson still pleaded for some
scheme of abolition, and tried to devise a plan for
deporting the blacks to a favorable clime. Just before
his death he wrote, " I leave this to those who will live
to see the accomplishment, and to enjoy a beatitude forbidden to my age. But I leave with the admonition

to be up and doing." So far it had been demonstrated that free labor and slave labor could not co-operate.

Migration had moved mainly on parallels, so that New England and Pennsylvania had taken chief possession of all north of the Ohio River; and Virginia with the Carolinas had created Tennessee and Kentucky. The admission of new States into the Union was watched with about equal solicitude and jealousy at the North and the South. The balance was sure to be broken, and one party was about as ready as the other to prophesy disunion as a consequence. The origin of opposition to the admission of Missouri with slavery, in 1819, was unquestionably with the old Federalisits, whose anti-slavery sentiments were deeply tinctured with political ambition. Jefferson, whose antipathy to slavery was stronger than their own, was intensely angered by the reappearance of Northern jealousy that had already been so ready for mischief. But the sentiment of anti-slavery was not confined to political schemers. It was a strong moral conviction growing daily more intense. As the Northern people swept slavery out of their section they were able to see the moral evil without selfish bias. After a great amount of preliminary skirmishing, in January of 1820, a bill which had been sent to the Senate from the House, to admit Maine into the Union, was reported with a clause admitting Missouri on the same unrestricted terms as to forms of labor. The great battle which was destined to last for forty years, intensifying sectionalism until cleavage occurred, now began. The two sides came up fully and frankly as North and South, free labor and slave labor. The whole country was growing heated. Meetings were held everywhere to express public sentiment. Memorials were sent up praying Congress to refuse to permit any extension of slavery. Women flocked into the Senate chamber until John Randolph drove them out with insulting sarcasm. The Quakers came to the front for freedom. Pennsylvania appealed to sister States to refuse to covenant with crime. Virginia retorted that the Constitution gave Congress no right to interfere with domestic institutions.

In Congress there was more than muttering, there were threats. Clay was puzzled to work in a compromise ; but prophesied three confederacies inside of five years. Pinckney, the most distinguished of Southern orators, was more brilliant than ever he had been in a nobler cause. John Quincy Adams foreboded war. President Monroe worked for a compromise. The Senate after a sharp fight adopted an amendment that no slavery should be allowed north and west of Missouri. The House was soon committed to the principle that Missouri should not be admitted without the prohibition of slave labor. But in the end a compromise measure was accepted by a large majority. Maine was admitted without stipulations, and Missouri on the same conditions. The Missouri Compromise was now a part of American history ; and like all compromises involving moral principle, promptly began its work of alienation, rather than fraternization. No one was quite satisfied. Most were fearful that they had only delayed a storm. Randolph cursed the " dirty bargain," and the " dough faces " from the North who had helped to enact it.

Slavery was now forever interdicted north of 36° 30'. It was not likely to go there, if not debarred. But south of that line it could find its cordial home. Missouri by the compromise received only an enabling act. A convention of her people at once formed a constitution strongly pro-slavery in character, and excluding free negroes from living in the State.

The discussion concerning her admission under this Constitution was fully as bitter as any that occurred from 1850 to 1860. It was as unexpected as it was threatening, for no one had fully appreciated the change of sentiment going on in the South. Mr. Pinckney said, " This is the one question which may in its consequences lead to the dissolution of the Union, and prove the deathblow to all our political happiness and national importance." Mr. Floyd of Virginia in the heat of the strife said : " Sir ! we cannot take another step, without hurling this government into the gulf of destruction. For one I say I have gone as far as I can go in the way of compromise ; and if there is to be a compromise beyond that point it must be at the edge of the sword." The necessity of counting electoral votes occurred in the midst of this wrangling. The joint meeting of the Houses was riotous. Objections were made to receiving the vote of Missouri. The Senate in anger rose and left the hall. The tumult was too great for debate. Some cried, " Missouri is not a State " ; others were crying, " Missouri is a State." Tallmadge, who represented the anti-slavery forces, said : " Has it already come to this, that in the legislative councils of the American Republic the subject of slavery has become the subject of so much feeling, of such delicacy, of such danger, that it cannot safely be discussed ? Are we to be told of the dissolution of the Union, of civil wars, and of seas of blood ? And yet with such awful threatenings before us, do gentlemen in the same breath insist upon the encouragement of this evil—upon the extension of this monstrous scourge of the human race ? If its power,

its influence, and its impending dangers, have already
arrived at such a point, what will be the result when it
is spread through your widely extended domain ?   Its
present threatening aspect and the violence of its supporters, so far from inducing me to yield to its progress,
prompt me to resist its march.   Now is the time.   It
must be met now ; or the occasion is irrevocably lost,
and the evil can never be controlled."   King of New
York fully measured the position when he said as a
philanthropist, " Freedom and slavery are the parties
which stand this day before the Senate ; and upon its
decision the empire of the one or the other will be established."   As a statesman he added that the Constitutional provision allowing a representation by proxy for
slave chattels had given a " preponderance to the slaveholding States over the other States of the Union.
Nevertheless it is an ancient settlement, and faith and
honor stand pledged not to disturb it.   But the extension of this disproportionate power to the new States
would be unjust and odious.   The States whose power
would be abridged and whose burdens would be increased by the measure can not be expected to consent
to it ; and we may hope that the other States are too
magnanimous to insist on it."   " The discussion,"
says John Quincy Adams, " disclosed a secret.   It
revealed the basis for a new organization of parties."
Finally Clay " the peacemaker " secured the reference
of the whole matter to a compromise committee of thirteen, who reported that the agreement of the previous
session should not be disturbed ; and that Missouri
should be admitted on the simple condition that her
constitution should not be interpreted to the detriment
of citizens from another State.   This compromise of a
compromise was a subterfuge, of no value whatever as

far as the main subject was concerned. But it threw
the subject out of the House. Missouri made a nominal
assent to the requirement of Congress ; and Monroe by
proclamation announced that a new star was to be
added to the flag.

So far as slavery was concerned the two sections were
now on reasonably kindly terms. But slavery had
begun to eat the life out of Virginia and most of the
rest of the South, while free labor by every stroke sent
the North ahead. Free education for every brain and
unshamed toil for every arm shortly made a terrible
contrast with a section where one-third of the people
must be kept in ignorance in order to prevent insurrection, and where the other two-thirds were ashamed to
use their brains on inventions or their hands at honest
gains. At the North the poorest were encouraged to
grow rich ; at the South a poor-white class was created,
to a permanent inheritance of servility, conceit, and
ignorance. Jefferson, with all the energy of his declin-
ing years, struggled to reconstruct society on a safer
basis. He had his slaves organized as an economist
establishment, and taught trades. He labored most
strenuously to create a system of higher education ; but
before he could get his State university chartered in
Virginia Michigan stole his idea ; and the whole free
Northwest, already having New England common
schools, began to cap them with Jefferson's university
system. It is busy to-day correlating the two. While
Virginia was waiting for roads to be built for her across
the ridges to her colonies, New York, alone, dug her
Erie Canal ; and set the water of the great lakes to kiss
the ocean in her harbor.

Only a lull in the storm had been secured. Great
Britain meanwhile had joined with the United States to

suppress slave-trading in Africa and on the seas. Li-
beria was established, and men of both sections were
looking yearningly for some way of getting rid of the
problem of a negro population, by deportation.   Naturally the South, planning for some means of repudi-
ating unfriendly legislation, held strongly to the
independence of States ; the North no longer cared for
this doctrine, and rarely appealed to it after 1820.   But
the sectional divergence was most essentially one of
social character.   The question was not to be escaped
whether this divergence would make it impossible to
sustain political unity.   De Tocqueville prophesied a
necessary separation.   There was at that time nothing
to show that the vast expansion of the Republic would
introduce other questions, on other lines of cleavage,
and that one form of selfishness would balance another.
But we see now that the larger a Federal Union the
safer ; and that had slavery in 1820 occupied only the
southeast corner of an ocean-bordered union, as it did
forty years later, it would never have caused just
alarm.

Probably no effort at pacification was ever more unfortunate than this which, as a Compromise of sections,
selected a definite geographical line, and erected it into
a perpetual barrier.   Jefferson, always our greatest
mind, instantly saw the fateful blunder.   " I considered
it," he wrote, " as the knell of union.   The coincidence
of a marked principle, moral and political, with a
geographical line, will I fear never be obliterated—
renewing irritations until it kindles such moral hatred
that separation would be preferable " ; adding almost
the only pessimistic passage in all his writings, " I
regret that I am now to die in the belief that the useless sacrifice of themselves by the generation of 1776 to

acquire self-government, and happiness to their country, is to be thrown away by the unwise and unworthy passions of their sons ; and that my only consolation is to be that I shall live not to weep over it.''

The danger of division was all the greater from the fact that steamboats were in their infancy, and railroads were practically unknown. Boston was five days from Washington by speediest post. The day that John Adams died in 1826, his son, the President, held a public levee in honor of the fiftieth anniversary of Independence. Two days later he heard of the death of Jefferson at Monticello, but not till the ninth of July did he learn that his father and Jefferson, at the same moment, five days before, had ascended from the scene of their sublime work as nation builders. Adams dreading most of all to leave his dear Republic, whispered as he went, '' But Thomas Jefferson survives.'' As if he would have said, I can be spared if only my great friend remains to watch and counsel. His early friendship for Jefferson had revived and become a passionate faith and love.

While the nullification acts of 1832 are in no way traceable directly to slavery, it is true that the drawing of a sectional geographical line immensely hastened the formation of sectional tastes and sentiments. In 1820 Calhoun was a Unionist, as strongly pledged to federal perpetuity as was John Quincy Adams. In the cabinet of Monroe he had voted that Congress could exclude slavery from national territory. Monroe loved him with a tenderness seldom manifested. Adams, who rarely showed affection through the opacity of his sternness, was warm in his admiration of the brilliant, chaste, incorruptible, boyish Calhoun. But a few years later this same rich, rare, beautiful soul; broad, liberal, generous, tolerant, declared in the Senate, that he " would not turn on his heel to be President ; that he had given up all for his own brave magnanimous little State of South Carolina." Miss Martineau wrote : " His mind has long lost all power of communicating with any other. I know of no man who lives in such utter intellectual solitude. Mr. Calhoun is full of his nullification doctrines."

However, the events of 1832, when the earliest fruits of sectionalism began to be reaped as nullification, we must first approach from another direction. Slavery drew the line ; everything else adjusted itself to that line.

The country never gained in wealth faster nor was prosperity more general than from 1800 to 1805. The country was pre-eminently agricultural. About ninety per cent. of the people tilled the land. It is true that these farmers were also manufacturers. They made their own shoes and stockings, and wove the cloth that clothed them, from fleeces of their own growing. They made their own candles, and soap ; and often shod their own horses. They built their own houses and school-houses, churches and mills. Cotton they knew little about, but woollen and flax they could manufacture for all domestic uses. For all this they neither asked nor had any protection. Washington, Jefferson, and Madison escaped from office as soon as possible to imitate Cincinnatus. Jefferson insisted that the permanent progress and freedom of the Republic depended on its remaining peculiarly agricultural.

The question of encouraging manufactures separately from agriculture had, it is true, been entered upon as early as 1789. Madison urgently argued that the tariff of that date was false in general principles and provoca-

tive of local antagonisms.   But the argument, since so
well known, was plausible.   " (1) We must have
revenue.   (2) Manufactures already established must
not be allowed to go down.   (3) We are liable to
future wars, during which we must be independent of
outside nations for our goods.''   The opposition showed
that the South was not interested in the protection of
manufactures ; and that such a policy would lay a burden on the agriculture of that section, for the advantage
of the enterprises of another section.

But the pressing question was immediate revenue ;
so that concession and compromise were not difficult to
bring about.   Madison said, " I own myself a friend
to a very free system of commerce ; and hold that commercial shackles are generally unjust, offensive and impolitic.''   He had the policy of England and Napoleon
fresh in mind.   " It is also a truth that if industry and
labor are left to take their own course they will generally be directed to those objects which are the most
productive; and this in a more certain and direct manner than the wisdom of the most enlightened legislation
could point out.''   He however yielded that absolute
freedom of trade must be a matter of general reciprocity.   " There are cases in which it is impossible to
avoid following the example of other nations.''   The
debates showed the immense power which protection
ever has to stir up all the selfishness of human nature.
One wanted a duty on candles, another on rum.   One
desired to prohibit foreign beer.   Steel was in its infancy, and was crying aloud for help.   Fisher Ames of
Massachusetts wanted nails encouraged, and Tucker
of Virginia strongly opposed.   The debate drifted into
a confession that commercial discrimination was the
cause of our war with England ; that we ought not to

enter on a deliberate system of commercial warfare. It was urged further that to impose a tariff on imports was only to tax indirectly a part of the people for the advantage of another and smaller part. But from the very outset the word " protection " served as a fortunate one for those who meant taxation ; a word hated as badly as protection was admired.

In 1787 Hamilton laid before Congress his elaborate and masterly argument in favor of sytematically building up manufactures. He argued that manufactures would not depress agriculture ; would encourage immigration ; would build up a home market of consumers. He would in fine diversify industry ; and he would do it by a tariff. Jefferson however declared the whole scheme to be adverse to liberty, and a machine for the corruption of the legislature. Already a powerful lobby was growing up, to work on Congress by corrupt means, in favor of local or sectional interests. Madison replied that he considered the scheme " as subverting the fundamental and characteristic principles of the government, as contrary to the true and fair as well as the received construction of the Constitution. Jefferson, on this antagonism to a system of taxation which he called infernal and eternal, built up the great Republican party; and in 1800 was elected by it to the Presidency.

The effect so far had been scarcely more than to increase the retaliation of other nations. Jefferson declared that instead of embarrassing commerce by a complexity of regulation duties and prohibitive laws, he would urge a general effort among civilized nations to remove all discriminations ; and let each country manufacture what it could do best, and what nature had appointed for it. He believed that such an attempt might wisely be made with only one co-operating nation. But the embargo system which he allowed his party to adopt was as hard a tax on New England as the protective tariffs were on the South. It was a hopeless struggle against the vicious commercial discriminations of both Great Britain and France. But that any other course would have steered us through the era of the Napoleonic wars with more safety, and better results, is not probable. Gallatin and Madison were compelled to waive their convictions to the exigencies of the times ; but they always held that protective tariffs must stand as war measures. Nations should not continue a vital struggle to destroy each other by legislation while nominally at peace. But there came about a curious and unthought of result of the embargo and the war of 1812. New England, stopped in her trading and importing ventures, turned all her wonderful enterprise to make what she could not buy. By the close of the war she was transformed into a manufacturing country. Her wares of every conceivable sort were sent southward in trains of wagons that were almost incessant. In 1815 there were one hundred and forty cotton manufactories in and near Providence, R. I. The number of cotton spindles in operation in the United States, which had increased from 3000 in 1798 to nearly 20,000 in 1808, was now estimated at 500,000 ; employing \$40,000,000 capital, 100,000 workmen, and paying \$15,000,000 wages yearly. New England had opposed the war almost to the limit of secession ; but to her own surprise she had waked up on the high road of unprecedented prosperity. She was transformed from a commercial to a manufacturing people. This was a war boom ; could it be sustained in times of peace ?

When the war closed she faced ruin. Opening the ports would let in foreign goods once more and ruin

manufactures.   As soon as peace was declared, Dallas
was called upon by Congress to report a new tariff
measure.   The pressure was tremendous.   Sectional
opposition was tolerant and generous.   Calhoun,
Lowndes, and other Southern leaders admitted the
desirability of preserving what manufactures were
already established.   Protection was modest in its
demands, and the self-interest of manufacturers was
restrained.   Clay, although a moderate protectionist,
pictured in glowing colors the old ideal of a family that
did its own manufacturing as well as soil tillage.
Randolph almost alone fought the doctrine of protection as a false principle from bottom to top.   He said :
" I will buy where I can buy cheaper.   I will not
agree to lay a duty on the cultivators of the soil to encourage manufacturers."   Even without the aid of
Congress, the agriculturist, he urged, was no match
for the manufacturer.   " Alert, vigilant, enterprising,
and active, the manufacturing interests are collected in
masses, and ready to associate at a moment's warning,
for any interest to their body."   Already the percentage of agriculturists was rapidly running down.   It
was urged that while manufacturers were coming
forward to be helped, nothing was said of the losses
of other classes by the war.   Agriculture must stand
on its own feet, bear its own burdens ; and besides that
be taxed to sustain a rival industry.   Telfair argued
strongly, " Because monopolies have for ages become
familiarized to us, are we to disregard the evidence in
favor of an unshackled pursuit of our own interest ;
and in despite of the warning voice of these very
nations, which attests the ruinous effects of such a
policy, upon every principle held sacred by the friends
of freedom, are we to give aid to a favored class of the
community by a tax on the rest ? "

Clearly already the country was entering on a course that would soon be intolerable to agriculture, and bring about a sectional clash. Not only so, but those monopolies that have outlived the sectional age, and become the curse of a later period were set on foot and protected. Momentum was given to concentring population, and stripping the farms of the most enterprising youth, while wealth tended steadily to accumulate in fewer hands. Webster, who began as a free trader, had gone with New England over to protection ; but he declared he was " not ready to accelerate the approach of the period when the great mass of American labor shall not find its employment in the fields, when the young men of the country shall be obliged to shut their eyes to external nature in the heavens and on the earth, and immure themselves in close and unwholesome workshops ; when they shall be obliged to shut their ears to the bleatings of their own flocks upon their own hills, that they may open them in dust and smoke and steam—to the perpetual whirr of spools and spindles and the grating of rasps and saws." He urged that government should allow all the different pursuits equality.

Hamilton at this period began to be " One of the brightest stars in our political hemisphere ! " He was continually quoted by protection advocates, while Jefferson was sneered at, and Madison ridiculed. Ingham, who was afterwards in Jackson's cabinet, but turned out of it again in one of that chieftain's pell-mell revolutions, attacked Madison's theories as fallacious. Protection was, in his judgment, the only great national principle. Calhoun was by no means a foe to manufactures. He consented, as we have seen, to the tariff of 1816, on the ground of a reasonable balance of industries. But he strongly urged that we must not undertake to keep up enterprises that had been brought into life only by war. He said, " On general principles a certain encouragement ought to be extended at least to our woollen and cotton industries." The tariff of 1816, however, totally failed of doing what was intended by it. It was to protect manufactures that existed. It was discovered, however, that people must go deeper, and look more broadly to protect any industry. (1) Manufactures had expanded beyond possible demand. (2) Fashion changed, and a rage arose for foreign goods, which took the market from better home goods. (3) The boom and inflation fostered by protection, encouraged bad business methods. The crash came. Agriculture, without protection, was moderately prosperous until 1819–20. But the mischief was working. "Thousands of persons forsook their farms." Wages of farm laborers went up so as to embarrass agriculturists for lack of sufficient help. Commerce was suffering from the instability of exports.

The demand arose from all manufacturing quarters for *more* protection ; higher tariff rates. " Russia," said their Apostle Carey, " is a hundred years ahead of the United States, for she positively prohibits, under penalty of confiscation, all articles her own people can manufacture." He forgot to say that only a slightly more exclusive policy was furnished by China or Japan. Higher and higher must go the rates ; and little was said about the " infants " ever becoming able to walk alone. Evidently protection was becoming a fixed principle. Indeed, how should a people ever be able to get rid of a system that appealed to every selfish instinct, and blinded men to the common and equal rights of their neighbors ? Henry Clay, who still believed himself to be a free trader, argued with his peculiar eloquence, that such an ideal could not be thought of without universal reciprocity.

With Carey and Niles he began, out of the transient protective principle, to create a new philosophy of economics. This alarmed those who heretofore yielded kindly to the demands of manufactures. Clay had baptized this policy " The American System." The North was crowding. The South took alarm. It was said that on this plan, pushed far enough, the farmers must become manufacturers, or quit the country. Already the percentage of agriculturists had gone down from 96 per cent. to nearly 70 per cent. What was to be the outcome ?

The Tariff Bill of 1820 was lost by one vote. Many, like Otis and Silsbee, were not ready to enter on a fixed principle of " building one interest on the ruins of another." Tyler urged that, if let alone, capital would flow into such channels of employment as gave it best returns. In other words, we were doing the worst possible thing for manufactures to place them on a false and artificial basis. Commerce protested as loudly as Agriculture. A memorial from Charleston demanded that labor and capital should not be artificially controlled ; but that, if the Northern cities were to be built up by protected manufactories, Southern cities should have protected houses to raise tea, pepper, etc. Memorials had heretofore been mostly from those seeking protection. It had been left to them to bring forth new arguments. But now it was shown that Great Britain would be glad enough to get out of her protective system if she could. It was also proven that the Hanse towns, with totally unrestricted commerce, were prosperous beyond example ; and had outsold Great Britain with her almost prohibitive

system. Protection it was said was driving population
about in herds. " England was prosperous in spite of
tariffs.''

The failure of the Tariff bill of 1820 was followed by
great business prosperity. Farming and commerce
were convalescing. Business adjusted itself to the old
tariff. Unfortunately, Monroe, while congratulating
the country that " the history of the world furnished
no example of a progress" like ours, yet recommended
a farther stimulus to manufactures. In 1824 a Bill,
swathed neatly and wrapped in the folds of " The
American System," was enacted. It is amusing to a
student of history, to find Clay appealing to " that
master spirit of the age Napoleon Bonaparte " as the
highest authority on the question. We are now able
to read Napoleon's personal exposition of his unmitigated selfishness in his conversation at St. Helena.
Webster declared the American System was no such
thing ; but that it was an effort to foist, under cover of
that name, a purely foreign and despotic system on
the American people. To Wm. Giles, Jefferson wrote
December, 1825, " Under the power to regulate commerce, the government assumes indefinitely that also
over agriculture and manufactures ; and calls it regulation to take the earnings of one of these branches of
industry, and that too the most depressed, and put
them into the pockets of the other the most flourishing
of all.''

The battle in 1820, coming in close connection with
the contest over slavery, intensified sectional feeling.
The tariff discussion in 1824 was not particularly bitter;
yet the administration of Adams was marked for its
strong centralizing tendencies. This emphasized the
State-rights sentiment at the South, and shaped the

contest between the two sections. State rights, which in 1814 had been a favorite New England ideal, was henceforth to be a Southern theory ; while New England swung back to federalism and a strong government. The only direct result of the tariff of 1824 was to prove the inefficiency of any tariff to protect, short of positive prohibition. The wool manufacturers complained that they were still undersold by England ; and that nothing would do but a " square yard duty," and a minimum rate. Capital, following exactly the same route as in 1820 and 1816, had gone headlong into over-production, so as to glut the market, even if England were barred out. Half the machinery had to be stopped. England meanwhile was not idle to her own interests, and removed her duty on raw material.

Of course the " American System " was not to be given up. The only remedy was once more higher rates. Raw wool was to be advanced to thirty-five per cent. after June 1, 1828 ; to forty per cent. one year later ; and wool costing between ten cents and forty cents per pound was to be listed as costing forty cents. This proposition stirred up a demand for raised duties all along the line. Nearly everything began to cry for protection. Capital was drifting into fewer hands, and was better able to influence legislation. Political conventions now took up the role of making tariff schedules. Congress was to be urged to accept of such measures purely as party affairs. The bill of 1828 was so offensive to wool manufacturers that they named it " the tariff abomination." It was hoped that by amendment it could be made so objectionable as to defeat it. But it passed by a decided vote. Although there were Southern protectionists, the sentiment of that section was overwhelmingly for freedom of trade, or at most a tariff for revenue.

The preliminary notes of indignant protest passed into vigorous and sometimes threatening resolutions. President Cooper of South Carolina College, asked at an anti-tariff meeting, "Is it worth our while to continue this Union of States, where the North demand to be our master?" A famous letter known as The South Carolina Circular declared : " If we do not at once seize upon the strong ground of principle, with the determination never to quit it, our cause is lost. Protection was never meant to become a permanent tax on the consumer, but to give a start to a new undertaking for a few years. Are our domestic manufactures to continue in perpetual infancy? Our national pact is broken." Historically this was unquestionably correct. South Carolina had been as generous as any State in sacrificing her specific interest for the general good. She now could fairly ask : "Are we to exist in the Union merely as an object of taxation? What is to be the end of this American System?" McDuffie, at a public dinner, voiced public sentiment, and was not far from the truth, when he said : " The protective system has ruined South Carolina commerce, depressed her agriculture and made bankrupts of her wealthy citizens. . . . Two-thirds of Congress are actuated by selfish ambitions and avaricious motives," and would pursue " their reckless course in spite of the ruin of that portion of the Union which produced more than two-thirds of the exports of the country." This was unfortunately the most dangerous phase of tariff legislation : that it blinded legislators to America, and intensified their local sentiments. McDuffie proposed retaliatory measures in the way of a heavy tax on Northern manufactures—an unconstitutional measure.

Such protests soon went farther and became positive demands and threats. The South Carolina *Mercury* proposed a State Convention ; and that the convention should send an ultimatum to Washington. Governor Hamilton declared that " the South had been drugged by the slow poison of the miserable empiricism of a prohibitive system " and that it was now impossible for the political parties to untangle themselves from the consequences. Congressmen must legislate to win votes. He insisted that the remedy and the only remedy was the assertion of State rights. He quoted the Resolutions of '98. " The several States who formed the Constitution being sovereign, and independent, have the unquestionable right to judge its infractions ; and a nullification by those sovereigns of all unauthorized acts is the rightful remedy." But there were other leaders not yet ready for extreme or questionable measures. State tariffs, which had been suggested, were clearly unconstitutional ; and of the same essential character as the evil that was combatted. It was agreed to foster home manufactures and economy. Governor Iredell of North Carolina and Governor Johnson of Louisiana frowned on any assertion of secession principles. Calhoun counseled moderation. Crawford of Georgia, who was at the height of his influence, opposed nullification. Jackson, who was running for President, it was expected would urge tariff reduction. But he was a politician and a soldier with an autocratic will, and determined on only one thing—to govern. His sympathies were naturally with federalistic methods ; but for republican measures. Nor had he been long in office before he found

Calhoun in his way. He turned all of the Calhoun men out of his cabinet, and by other adroit measures divided the South, so as to leave South Carolina largely isolated in its decisive sentiments for positive action. April 13, 1830, at a banquet on Jefferson's birthday, Jackson arose and propounded this toast, " Our Federal Union, it must be preserved." His speech was characteristic. It was done with all his bluff inconsiderateness, but it served the purpose to check threats of secession and nullification.

In December of 1831 Jackson's message congratulated the country on its prosperity, and the prospect of a speedy extinction of the national debt. There was no longer need of a high tariff for revenue, he therefore advocated a reduction of rates. This was to place the tariff as a revenue measure. The protectionists saw that it was vital to establish protection as an economic principle. The battle guage was instantly taken up. The debate was hot and often furious, both in and out of Congress. The Carolinas with Virginia and Georgia were the core of the free-trade section. Clay still held Kentucky to his own position of tariff for " limited protection." Matthew Carey lamented, as one of the most alarming features of the controversy, that " a large portion of the most decided supporters of the Union, and enemies of nullification, and its counterpart the dissolution of the Union, with all its attendant horrors, are firm believers in the unconstitutionality of the protective system, and appear to require its total abolition." Monster meetings were held in all the large cities advocating each side of the question. Free traders were by no means scarce at the North ; and as a rule they were an intelligent class; who also looked with dread on the increasing centralization of power, and the extension of legislation by Congress. Albert Gallatin argued with great ability against protection as a false principle in economics; and it had not yet been forgotten that Gallatin had in 1800 reversed the drift toward national debt, and inaugurated an era of prosperity.

In Congress the struggle was not only momentous in import, but the ablest in champions that body had so far ever known. Henry Clay believed in compromise; it was his first and his second nature. The doctrine of protection as he held it, the American System, he believed would win both sections, and make him President. Calhoun, browbeaten by Crawford, was steadily being crowded back upon his own State. Webster had slowly become an extreme protectionist. McDuffie introduced a bill to reduce duties by degrees to 12½ per cent. This was the Southern ultimatum. McLane introduced a bill for a much less thorough and sweeping reduction. Adams was begged to act as umpire, and try by a just and judicious measure to pacify all parties. His bill finally became law. But like nearly all tariff legislation it created only increased dissatisfaction. Discussion was no longer possible. Nullification was the watchword of a united South. A seven-striped flag was run up in Georgia. South Carolina assumed its sovereign rights, and refused by act of legislature to allow the tariff to operate inside its borders. Jackson threatened to hang Calhoun; and warned South Carolina that the laws of the Union should be enforced in her limits.

The real aim and intent of the nullifiers of 1798 we have seen to have been to recall Congress to constitutional line. The nullifiers of 1832 began with the same spirit. They announced their love for the Union, and deprecated any alienation. A Union party was formed to resist extremists. Among the rest were Lowndes, and Hugh Legare, and Benjamin Hunt, and Harleston Read, and David Williams. The last-named said with force, "We were not sparing of our censures when New England meditated resistance to the Embargo. There was not a man among us who did not pronounce the Hartford Convention a traitorous association. It becomes us to look well to it that we do not tread in the very footsteps we have denounced." But these very men drifted in time to become pronounced nullifiers, at all costs. Calhoun never avowed himself unfriendly to the Union. Governor Hamilton issued a proclamation for a day of fasting and prayer for the removal of oppression. But the Charleston *Gazette* said : "This everlasting cant about devotion to the Union, accompanied by recommendation to acts that must destroy it, is beyond endurance. We know only two ways under our government to get rid of obnoxious legislation. We must convince a majority of the nation that a given enactment is wrong, and have it repealed in the form prescribed by the Constitution, or we must resist it extra-constitutionally by the sword." The road was open and unmistakable. The majority was placing itself where it must resist by the sword. It was charged by the Unionists as late as 1832 that emissaries had gone to England to see if assistance could be obtained in case of war. Petigru and Dayton kept up the fight bravely against the tide. Thomas Grimke even met the nullification act with a counter-blast that told a volume of prophetic truth. He said, " The ordinance passed by your convention is the grave, not the bridal chamber of liberty. It will be so regarded even in the South Carolina of future years, with grief

and mortification. The world may be called to gaze on
the blockade of your coast ; on the alternate execution
of traitors to the State and traitors to the Union ; on
the battle-field of brothers and the conflagration of your
towns. But to that world it will be the history of a
rebellious province, not of an independent nation."
A convention of Unionists followed. Randall Hunt
offered this resolution, "That the Union party acknowledges no allegiance to any government except that
of the United States." In view of the fact that nullifiers were perfecting a military organization, it was
inquired if it be not necessary for the Unionists to do
the same. Resolutions were passed denouncing nullification as practically secession ; and more tyrannical
than tariffs. Had Congress not yielded and brought
about pacification, the people must certainly soon have
drifted into war. The Unionists meanwhile would
have been the first, and possibly the only victims.
There are chapters of history that are never written but by the pens of prophets. It is well that it
is so. With Jackson for President, and his Florida
record to judge from, and his hate for Calhoun as an
incentive, it is almost certain that no measures of con-
ciliation would have been considered, whenever he
should have had control of the affair. He had already
ordered General Scott to South Carolina ; had collected
troops, and ordered two war vessels to Charleston. His
proclamation was not written by himself but by Edward
Livingston, a man of very different temper. It was
well for all parties. Its moderation and logical firmness being attributed to Jackson, made him at once a
national hero. So eager were the people for this State
paper, that a thousand copies were thrown from the
windows of the *Globe* office, before the mob about the

doors was sufficiently thinned to permit of ordinary
business. A force bill followed, empowering the President to collect the revenue in South Carolina with
United States troops.

But the proclamation on some points was sharply
questioned. It laid great emphasis upon and seemed
to accept the Websterian theory that the Union was
formed by the people and not by the States. "The
unity of our political character," it said, "commenced with the very existence of the government.
Under the Royal government we had no separate character; our opposition to its oppressions began as the
United Colonies." Historically this was untrue. It
and more like it was apparently in contradiction to
other parts of the proclamation. Had two heads been
at work; or two wills shaped the document?

The whole nation was in a ferment. The Republican
party charged the President with reviving old Federalism, and advocating centralized government. A resolution of condemnation was introduced in the Virginia
legislature to this effect, "That President Jackson has
set forth in the late proclamation that the Federal Constitution results from the people in the aggregate, and
not from the States." The resolution affirmed that
"This theory of our government would tend in practice
to the most disastrous consequences, giving a minority
of States, having a majority of population, the control
over the other States." The Richmond *Enquirer*
hastened to give Jackson's explanation, "authoritatively"; which was, in the best view to be taken of it,
a back out. After a very tedious effort to adjust himself to both positions, he flatly concludes. "No part of
the proclamation was meant to countenance principles
ascribed to it. On the contrary, its doctrines if con-

strued in the sense they were intended, and carried out, inculcate that the Constitution of the United States is founded on compact ; that this compact derives its obligation from agreement, entered into by the people of each of the States, in their political capacity, with the people of the other States. That in the case of a violation of the Constitution, and the usurpation of powers not granted by it, on the part of the functionaries of the General government, the State governments have a right to interpose, and arrest the evil ; upon the principles which were set forth in the Virginia resolution of 1798 ; and finally that, in extreme cases of oppression, (every mode of Constitutional redress having been sought in vain) the right resides with the people of the several States to organize resistance against such oppression—confiding in a good cause, the favor of heaven, and the spirit of freemen to vindicate the right.''

It was well known that Livingston and not Jackson was author of the proclamation ; and it was of nearly equal interest to determine precisely his views on the relation of the States to the nation. Mr. Livingston's views in summary I shall give in the Appendix to this chapter. But the following quotations are from a speech made by him in the Senate : '' The States existed before the Constitution ; they parted only with such powers as are specified in that instrument ; they continue still to exist with all the powers they have not ceded ; and the present government itself would never have gone into operation had not the States in their political capacity consented. That consent is a compact ; it is a compact by which the people of each State have consented to take from their own legislatures some of the powers they had conferred upon them, and to transfer these with other enumerated powers to the government of the United States, created by that compact. Although the Supreme Court must judge of the constitutionality of laws, and its decrees must be final, I am far from thinking this Court is created an umpire to judge between the General and State governments. In an extreme case an injured State would have a right at once to declare that it would no longer be bound by a compact which had been grossly violated."

So it appeared that when the issue came to a quiet and calm judgment, South Carolina and Jackson, including Livingston, did not disagree as to the general principle of " Compact "; and the right of a State or States to resist oppression. They only disagreed as to what constituted oppression ; and what should be the order and method of resistance. Was South Carolina oppressed ; and had she taken a wise and Constitutional course to secure redress ? This in fact must be the question in every case of disagreement as touching legislation inside the nation. The course of New England leaders in 1803 was pure treason for party purposes. Aaron Burr contemplated nothing short of treason. The Hartford Convention proposed to break up the Union for reasons that no one now looks upon as substantial. Was South Carolina in the same category ? Jackson was certainly right in saying that the purpose of staying in the Union and obeying what laws she chose was absurd. The proclamation granted that a great grievance existed ; but asserted that this was about to be greatly mollified. South Carolina had not asked " For a convention of the States to consider her complaints."

To this South Carolina answered with defiance. Governor Hayne issued a proclamation with this whereas, " The President of the United States has issued his proclamation denouncing the proceedings of this State, calling upon the citizens thereof to renounce their primary allegiance, and threatening them with military coercion, unwarranted by the Constitution, and utterly inconsistent with the existence of a free State ''; followed by a discussion of nullification as a rightful remedy for oppressive legislation. But the gist of the whole matter was that South Carolina had undertaken a '' peaceful revolution ''; and Jackson was right that revolution and peace were not coincident terms. If she chose to revolutionize she must bravely take the consequences. This she could not avoid by declaring coercion to be unconstitutional. She had deliberately written an address to the other States, saying that her '' withdrawal '' would doubtless '' end in the dissolution of the Union.'' Yet she denied the right and Constitutionality of any steps to coerce her; while to secure her ends she would coerce all the rest of the States. In plain English, her course was coercive from the beginning. She must expect to meet with coercion. Whoever appeals to force must abide by the consequences.

With the very thick glove of Livingston there is, nevertheless, clearly felt in the proclamation Jackson's iron hand. He does not hesitate to hurl the strong epithets of recklessness and madness. The people of South Carolina are plainly told they have exactly the rights of revolutionists—no more—and they need not whine about it. Peaceable secession is nonsense. He allows that her grievance is real, but not so great that she has an excuse for tearing down the Union.

The question of State sovereignty, as dwelt upon by Livingston, deserves special attention. He says, '' How then can that State be said to be sovereign and inde-

18

pendent whose citizens owe obedience to laws not made
by it, and whose magistrates are sworn to disregard
those laws when they come in conflict with those passed
by another." But he passes at once to discuss whether
the States have an undivided sovereignty. Here lies a
continent-wide distinction. The State retains sovereignty, but yields sovereignty. There is no logic in a
play upon words, as if sovereignty needfully meant
total power. It means supreme and original power so
far as its power is not given away. Stephens says very
justly in his " Constitutional View of the War between
the States," that " only sovereign States could endow
a common government with sovereignty." And as it
is allowed they created only a limited national government, the original and remaining sovereignty is still
their own. The point Livingston really aims at is not
to disallow sovereignty in the States, but to demonstrate
sovereignty in the nation. This may be granted at
once, that sovereign States granted sovereign powers ;
that they could in fact grant no others. The nation is
sovereign. The States are sovereign. But the nation
must not exercise its sovereignty beyond bounded constitutional limits. That is the whole meaning of the
Constitution. It limits national sovereignty ; and it
impliedly sustains State sovereignty, with limits also,
of their own creating for common good. The whole
point then comes back to this, does nullification of a
national statute come within the range of reserved State
sovereignty ? Here we must listen once more to Mr.
Livingston : " If, in creating a sovereign nation, a
sovereign power was given by sovereign States, whereby they yielded power to judge of national legislation,
the question is settled. Clearly nothing of the kind
was done. But a court was created to review Congres-

sional action, and, with great deliberation, pass upon the Constitutional quality of such action." But as has been shown in Chapter First, such a court has proved to be inconclusive. Its decisions have been challenged by Congresses, by Presidents, by State Legislatures, by State Courts. Its power to nullify hasty Congressional action, and to delay hasty popular action is easily conceded. In the Fugitive Slave cases the Supreme Court was decided to be wrong ; and its decisions were reversed by the popular conscience. This was done by States ; but by half the States. How then shall we consider the case where one State protests and desires to nullify ? The right is just as defensible as in the case of twenty States ; but the assertion of it involves revolution. The general compact is a compact of compromises. Compromise is the science of yielding privileges and rights for the general good. Precisely this was felt by South Carolina. And the logic of Andrew Jackson was " This you must yield for the common good, or for the common good we will compel you." Force bills are not unknown to our history.

On the 16th of January of the next year, 1833, Jackson sent to Congress a special message on the state of the Union. In this paper he announced the receipt of official papers from the Governor of South Carolina. That State had ordered its military forces to be ready at a moment's warning. It had defied his proclamation of warning. The message of Governor Hayne and the inaugural of Governor Hamilton he transmits. They bristle with treason. "They not only abrogate the acts of Congress, commonly called the tariff acts of 1828 and 1832 ; they sweep away at once every act imposing any duty on foreign merchandise." He regards the action as revolutionary ; and he calls on Congress for prompt

action. " The right of the people of a single State to
absolve themselves at will, and without the consent of
the other States, from their most solemn obligations,
and hazard the happiness of the millions composing the
Union, can not be acknowledged ; but that a State, or
any other great portion of the people, suffering under
long and intolerable oppression, and having tried all
Constitutional remedies, without the hope of redress,
may have a natural right, when their happiness can be
no otherwise secured, and when they can do so without
greater injury to others, to absolve themselves from
their obligations to the Government, and appeal to the
last resort, needs not be denied." The message closes
with the solemn affirmation that the Constitution is
supreme and the " Union indissoluble."

Suddenly without consultation with either side, Clay
came into Congress with a compromise bill. Introducing it, he said that he wished " to reduce the rate of
duties to that revenue standard for which our opponents
have so long contended." Held to be a father of protection, he deliberately abandoned it. The measure he
proposed was not a compromise, but a surrender. He
said that on one side the argument had been that the
tariff taxed one portion of the people for the advantage
of another ; that it was a system always climbing up
and never satisfied; and on the other side he confessed
that those who favored tariffs saw only instability and
uncertainty. " Before one set of books is fairly opened "
he said, " it becomes necessary to close them, and to
open a new set. Before a law can be tested by experience another is passed." This was a complete picture
of our entire tariff history. He coolly gave up the
whole principle of protection ; and proposed the gradual
scaling down of all duties each two years ; so that in

1841 the bottom limit should be reached at twenty per
cent. basis.   This was held to be about the revenue
standard.   Practically the tariff would be thus taken
out of politics.   And it was for twenty years so far out
that business was on a stable basis, and an era of prosperity was rarely disturbed.   Privately Clay added that
he wished to prevent Jackson from a chance of exercising his vengeful passions on Calhoun and South Carolina.   In his speech he said : " If there be any who want
civil war, who want to see the blood of any portion of
our countrymen spilt, I am not one of them ; I wish to
see war of no kind ; but above all I do not desire to see
a civil war.   I think South Carolina has been rash, in-
temperate, and greatly in the wrong ; but I do not wish
to disgrace her, nor any other member of this Union.
Has not the State of South Carolina been one of the
members of this Union in days that tried men's souls ?
If we had to go into a civil war with such a State how
would it terminate ?   I do not wish to see her degraded
as a member of the Confederacy.   As I stand before
my God I have looked beyond party, and regarded
only the vast interests of this united people."   Manufacturers protested with a roar of disapprobation.   At
Boston they resolved that a surrender of the principle
of protection was cowardly.   Several legislatures took
strong ground in opposition to compromise.   Clay's
friends plead with him not to throw away his chance
for the Presidency.   He answered with Roman sternness that he " would rather be right than be President."
Calhoun, accepting the proposal of Clay, assured Congress and the country that he had not shifted his
ground.   He was willing now, as he ever had been, to
lend a hand to manufactures that needed it ; but he
would not assent to protection as a principle to be per-

petuated, and to be unlimited. It is not alien to our
topic to say that Mr. Clay's compromise and the surrender of Protection did not paralyze any industry.
Our domestic exports increased before 1860 from eleven
millions to thirty millions per annum. Our tonnage
more than doubled. The country never saw before,
nor has it since seen, such a development of material
and intellectual life. Internal as well as external
vitality was astonishing. The railroad and telegraph
in this period became the chief glory of man's brain
and energy ; and at the close of this period commerce,
manufactures, and agriculture were admirably balanced
industries ; making the buying capacity of the people
equal to its producing capacity. Our commercial tonnage almost equaled that of England ; and agricultural
products differed but ten per cent. from those of manufactures.

The compromise of Mr. Clay at once allayed the ferment, and put an end to military preparations. So
disappeared the nullification movement of 1832. Two
principles were established : (1) The Union is indissoluble except by revolution ; (2) Any section of States,
or any State, may revolutionize ; it cannot otherwise
nullify. The way was cleared for Lincoln in 1861 to
say, " No State, upon its own mere motion, can lawfully get out of the Union ; resolves and ordinances to
that effect are legally void ; and acts of violence within
any State or States are insurrectionary or revolutionary
according to circumstances." The Federal Union was
demonstrated to be of such a nature that an appeal to
force was made improbable ; and an appeal to argument
eminently wise.

The tariff controversy up to 1832 was as we have seen
a struggle of sections. Those who continually crowded

for higher rates along the seaboard, never, except in the suggested action of South Carolina, proposed a tariff between the States.   Indeed the corner-stone of the Constitution was prohibition of any such measures. By some logic strictly their own property, protectionists could see the necessity of an impost along the line of the lakes, but not along the line of the Ohio River. That members of Congress were personally interested in measures did not disturb their consciences as voters.   " Producers who have in view the direct interest of a monied pursuit can contrive better for carrying their ends than the incoherent mass of consumers." They can also better organize political programmes and capture conventions.   So it came about that a section, once protected, was steadily gaining in political power. Meanwhile protection, preventing natural competition in certain fields, left it to rage in others.   The farmer's perishable fruits must be hastened to his nearest market, only to find that some other section, with more favoring climate, had got in ahead of him, and reduced his compensation nine-tenths.   To-day Western beef forces the Eastern farmer out of cattle raising ; Western corn forces him out of raising that cereal ; and if he turns to orcharding, New England apples are set down near his door at prices that make it unprofitable to harvest his own crops.   Protection offers him no palliative. Nor did, nor does, protection prevent competition of another sort.   Inside the tariff cordon it permits and encourages, it even makes necessary trusts and combines, that run prices up or down at will.   In fact tariffs have always so failed to protect, that where they operate, trusts must be formed to supplement their action.

The evils of the protective system were lately summed

up by Ambassador Bayard as a form of State socialism,
" which has done more to foster class legislation, and
create inequality of fortune, to corrupt public life, to
banish men of independent mind and character from
. the public councils, to lower the tone of national representation, blunt public conscience, create false standards
in the popular minds, to familiarize it with reliance
upon State aid and guardianship in private affairs,
divorce ethics from politics, and place politics upon the
low level of a mercenary scramble, than any other single
cause.   Step by step, and largely owing to the confusion of the civil strife, it has succeeded in obtaining
control of the sovereign power of taxation ; never hesitating at any alliance, or the resort to any combination,
that promised to assist its purpose of perverting public
taxation from its only true justification, and function
of creating revenue for the support of the government
of the whole people, into an engine for the selfish and
private profit of allied beneficiaries and combinations
called trusts.   Under its dictation individual enterprise
and independence have been oppressed, and the energy
of discovery and invention debilitated and discouraged.
Gradually the commercial marine of the United States
has disappeared from the high seas, with the loss of
the carrying trade, and the dispersion of the class of
trained seamen and skilled navigators."   Agriculture
and commerce alike are prostrated.   Instead of 96 per
cent. of agriculturists of one hundred years ago the
United States has now only 42 per cent. ; while our
marine is less by one-half what it was one hundred
years ago.   The South in this sectional strife was right
in principle.   Agriculture and commerce will hereafter
demand and at some time secure their substantial
rights.

In 1816 Calhoun had said, with the forecast of true statesmanship : " Let it be never forgotten that the extent of our Republic exposes us to the greatest of all calamities, to the loss of liberty, and even to that in its consequence, disunion. If we permit a low, sordid, selfish, sectional spirit to take possession of Congress this happy scene will vanish. We will divide ; and in the consequences of division will follow misery and despotism." Whatever were the errors of this great and good man, he was a statesman that our young men may once more learn to study and love. A character spotless as the blue sky over his native State, a heart free of guile, he hated tyranny, he abhorred selfishness, he foresaw the consequences of sectional legislation. Between nullification and corruption he would teach our young men to prefer the former.

That slavery led to the nullification act of 1832 is unprovable. But that it drew the geographical sectional line, we have already seen. It also remained as an irritating element in the struggle. Dallas replied to Hayne, " When the Senator asserts that slaves are too improvident, too incapable of that minute, constant attention, and persevering industry, which are essential to manufacturing establishments, he admits the defects in slave labor, he admits an inability to keep pace with the rest of the world." But Mr. Dallas forgot, and others with him forgot, that the interests of the whole Union as it was, were to be protected and fostered. He accepted the fact that the South could not compete with the North even without tariff, yet he demanded protection for the North, which he boasted was not disabled by its forms of labor.

Of the two great actors in this dramatic episode of American history, Jackson for the time being came out

the hero. Nor has history until very recently begun
to relegate him to his true position, and give a just
hearing to his antagonist. So far in our study of events
we have met Jackson in his attitude of alliance with
Burr. Even friendly critics have never been able to
acquit the fiery Tennessee general of an intention to
commit his State to a very questionable enterprise,
without consent of the general government. Again we
have seen him at the close of the war of 1812 the hero
of the magnificent battle of New Orleans. Arrogant,
presuming, arbitrary, uncontrollable, he was a natural
product of the times and conditions. We now find
him for the third time in our path. Determined to
secure his own re-election, he is ready to crush every
rival in his path. In some respects the greatest, in
others he is the meanest, in the roster of American
Presidents. He broke the peace with Spain by high-
handed military aggression. He was ready to join
Burr in fillibustering exploits that must endanger the
Union. He stole Texas, he broke up cabinets, and
defied Congress. He was a traitor to Monroe, and
more than false with Calhoun. He threw the finances
of the nation into confusion by arbitrarily seizing the
treasury ; and by despotic measures he elected a successor whom he expected would be subservient to his
will.

Calhoun was elected Vice President first with Adams,
and afterward with Jackson ; and fairly looked for the
Presidency. At this time he was an ardent nationalist.
At Augusta, Georgia, in 1825 he said, " No one would
reprobate more pointedly than myself any concerted
action between States for sectional purposes." But he
had not failed to consider such a possibility. He said
to Adams in 1820 that he did not believe slavery would

ever divide the Union ; but if so the South would be compelled to form an alliance offensive and defensive with Great Britain. Adams replied, " That would be returning to the colonial state."   " Yes," said Calhoun, " but it would be forced upon the South."   In 1828 he speaks of " our political system resting upon diversity of geographical interests."   When he went a little farther he fastened on the veto power of the States as the corner-stone of the Union. Webster became convinced in 1828 that " the idea of a Southern confederacy had been received with favor by a great many of the political men of the South."   So the secession ideas in which Webster drew his first political breath, had been transferred from New England to the farther extreme of the land.

The episode that finally hurled Calhoun back on South Carolina and sectionalism was his relation to Jackson's attempt to compel the women of Washington to admit to social equality Mrs. Eaton, wife of the Secretary of War. She had been charged with improper intimacy with Eaton before the death of her first husband. Very similar charges had been made against Jackson's wife. The President failed to even coerce the wives of his cabinet officers. But while Calhoun led in opposition, Van Buren bowed to his master. From that hour Jackson bent all his will to make Van Buren President, and crush Calhoun. He succeeded in both attempts. The disappointment of Calhoun was great ; for he alone of the crowd of Presidential aspirants had held himself honest and unpurchasable. Gradually he settled down to sectional statesmanship. He forsook the principle of the right of the majority to rule, to sustain the power and right of the minority to resist. Such champions however are not needless in a republic.

Calhoun was led unquestionably by laudable ambition for the Presidency. He had met with outrageous treatment ; himself too pure and upright to match the tactics of his rivals, the President, and Secretary of State ; he had been above reproach as a man and a statesman. His ideal of life was out of sight beyond that of Jackson and Clay and Crawford. But it had been his lot to be the victim of detestable plots and the malignity of his inferiors. Not a tolerable partizan ; a politician not to be mentioned in combat with Van Buren ; he was a statesman in the larger sense of the word, for he believed in principles. Nearer akin to Jefferson in views and character than any other public man of the era, he wholly lacked Jefferson's ability to widen with antagonism and expand with age. He believed in honesty, economy, and in liberty. His doctrine of State rights was the doctrine as it had stood in 1798. He drew back from nationalism ; he drew in from the combat at large. Under Van Buren he declared in the Senate his final stand. " The days of legislative and executive encroachments, of tariffs and surpluses, of bank and public debt, and extravagant expenditures are past for the present. The government stands in a position freer to choose its course than at any time from the commencement. We are about to take a fresh start. I move off under the State rights banner. I seize the opportunity thoroughly to reform the government, to bring it back to its original principles, to retrench and economize. I shall oppose strenuously all attempts to originate a new debt ; to create a national bank ; to re-unite the political and money powers, more dangerous than that of church and state ; and so prevent disturbances of the compromise which is gradually removing the last vestige of the tariff system.

Mainly I shall use my best effort to give ascendency to the great conservative principle of State sovereignty over the dangerous and despotic doctrine of consolidation." Here was Jefferson as he stood in 1798 ; but Jefferson was capable of adjusting his views to experience ; Calhoun was not. He would yield nothing, bend nowhere. Of necessity he dropped down to State leadership, and State rights, and State institutions. To become the champion of slavery because slavery was a State institution, was the final lot of a man apparently born to be the greatest national leader of his era.

But too much is always made of Calhoun in this issue of South Carolina with Congress. No Southern State compared with South Carolina for the intelligence and pride of character of its people. They were Puritans inside the established church. Their manners and habits were religiously moral. Their lights went out at nine o'clock. They believed in liberty as intensely as Sam Adams and the Bostonians. They were an inflexible stock. They are that to-day. Calhoun must be understood from this standpoint. The State stamp was a deep and indelible one. It did not breed nationalists, but it bred its strongest characters to look inward instead of outward. Nullification was a natural consequense. Its association with slavery was an incident, not a provoking cause. The deep cause of her action in two rebellions was State character.

South Carolina and Massachusetts from the outset, as they were in some senses most alike, were also most antipodal. Boston and Charleston stood steadily in sharp contrast. While Boston became the Athens, Charleston became the Edinburgh. Its population, highly aristocratic and conservative, allowed no invasion of morals or manners. When the Democratic convention of 1860 met there, Tammany sent its usual gang of claqueurs and bluffers. No other city in the Union but would have adjusted its social sentiment temporarily to these visitors. "Charleston simply shut its doors in their faces, with ill-concealed disgust. New York might indulge its vulgarity if it chose; Charleston would have none of it." The foreign element had barely touched the South. Its people, barring the mixture with Negroes, was more purely native than that of the North. The religion was Calvinistic and Presbyterian ; the politics Democratic, because all Charleston citizens were like those of Rome, held to be worthy of special privileges. But as for any bending to the notions and tastes of the cheaper herd, Charleston refused. State rights above popular rights were pre-eminently conserved. Its colonists were made up of Huguenots, and that of the very best blood of France ; Puritans who left England after the restoration of Charles II., dissatisfied with the loose morals of English life ; Protestant Irish; and some Scotch dissenters, with a few Swiss and Moravians. But before all was a large percentage of cavaliers of the English aristocracy. These finally fusing, created as proud and blooded a stock as America held. If Massachusetts was settled by a sifted residue of noble spirits, who disdained carnal things, South Carolina was settled by men who would sacrifice all for honor. We shall in time learn to turn to Charleston as we turn to Boston for national inspiration and social regeneration.

Calhoun was one in a long line of semi-monarchs ; for Carolina cavaliers never lost the royalty in their blood. They were the staunch believers in the Stuarts and Divine right. It is easy for a great Carolinian to

regard the worship of his State above the democratic
position of President of the United States. Calhoun
was no doubt entirely truthful when he said he preferred the love of his own State than to be the President ;
at least when he said it. Even at the close of the
century Charleston has not lost its sharply defined
characteristics. But New England has been made over
three times by immigration. Its theology is almost the
opposite of that which it held in 1800.

That the Supreme Court in 1832 constituted a safe
and honest arbiter against sectional legislation is a
vague fiction. As no department of the government
had been so factional in the factional era, so no department was so sectional in the sectional era. Jackson
had packed the Court with judges to carry out his
measures ; and it is well known that they did so ;
while he broke up the national bank ; defied Congress ;
stole the Treasury, and in general terms ruled as a
despot.

That nullification by any State, or by citizens of a
State, whether it be the nullification of a tariff ordinance or of a fugitive slave act is never defensible, need
not be here discussed. Certain it is that virtual nullification under cover of State ordinances, or State
court decision, or party platform, or by virtue of
" the higher law," has occurred throughout all our history. The final appeal has been to the people. Whether
the people will abide by the decision of the Supreme
Court, or the vote of Congress, will always remain a
question. That they will always wait until in the natural course of events legislative bodies can be induced
to reverse their acts, or courts revise their decisions is
not certain. Possibly our best escape from the dilemma
is to initiate, so soon as possible, the referendum, per-

mitting the people at once to veto objectionable statutes. But even then we shall not be freed from the friction of sectional interest. Our final hope is education of all the people, (1) never to crowd selfish legislation, (2) to be sufficiently patient, when such legislation is enacted, to try all constitutional measures before inaugurating revolution.

State nullification unfortunately proves too much. It involves the right of lesser sections to resist the State ; [1] and the end cannot be less than anarchy. If in South Carolina nullification had been inaugurated, there would have been not only outside military pressure but civil war within the State. The majority would have crushed the minority, and then the greater majority of the States would have crushed the victors. Republicanism is dependent on self-restraint, patience, and belief in the final honor of the majority.

[1] The tendency of extreme local sovereignty is well illustrated by the response of the sheriff at Moscow in Idaho to Governor McConnell. The latter had telegraphed that he would send troops to protect a murderer who was in danger of mob violence. The sheriff responded: " There is and has been no foundation for your statement. The sheriff's office will be conducted rigidly in accordance with law. The unbounded gall you exhibit in seeking to direct my office is no doubt surprising to those unacquainted with you. Obey the instructions you gave Grover Cleveland, ' Mind your own business,' hereafter; and keep your nose strictly out of my affairs." This was quite in the key of the reply made by the governor to the President when he proposed sending troops into the State to uphold law.

APPENDIX TO CHAPTER VI

JEFFERSON ON THE PROPOSITION OF CANNING

MONTICELLO, October 24, 1823.

DEAR SIR :

The question presented by the letters you have sent me, is the most momentous which has ever been offered to my contemplation since that of Independence. That made us a nation, this sets our compass, and points the course which we are to steer through the ocean of time opening on us. And never could we embark on it under circumstances more auspicious. Our first and fundamental maxim should be, never to entangle ourselves in the broils of Europe. Our second, never to suffer Europe to intermeddle with cis-Atlantic affairs. America, North and South, has a set of interests distinct from those of Europe, and peculiarly her own. She should therefore have a system of her own, separate and apart from that of Europe. While the last is laboring to become the domicile of despotism, our endeavor should surely be, to make our hemisphere that of freedom. One nation, most of all, could disturb us in this pursuit; she now offers to lead, aid, and accompany us in it. By acceding to her proposition, we detach her from the band of despots, bring her mighty weight into the scale of free government, and emancipate a continent at one stroke, which might otherwise linger long in doubt and difficulty. Great Britain is the nation which can do us the most harm of any one, or all on earth ; and with her on our side we need not fear the whole world. With her then, we should most sedulously cherish a cordial friendship ; and nothing would more tend to knit our affections than to be fight-

19

ing once more, side by side, in the same cause. Not
that I would purchase even her amity at the price of
taking part in her wars. But the war in which the
present proposition might engage us, should that be its
consequence, is not her war, but ours. Its object is to
introduce and establish the American system, of keeping out of our land all foreign powers, of never permitting those of Europe to intermeddle with the affairs of
our nations. It is to maintain our own principle, not
to depart from it. And if, to facilitate this, we can
effect a division in the body of European powers, and
draw over to our side its most powerful member, surely
we should do it. But I am clearly of Mr. Canning's
opinion that it will prevent instead of provoking war.
With Great Britain withdrawn from their scale, and
shifted into that of our two continents, all Europe combined would not undertake such a war. For how
would they propose to get at either enemy without
superior fleets? Nor is the occasion to be slighted
which this proposition offers, of declaring our protest
against the atrocious violations of the rights of nations,
by the interference of any one in the internal affairs of
another ; so flagitiously begun by Bonaparte, and now
continued by the equally lawless Alliance, calling itself
Holy.

But we have first to ask ourselves a question. Do
we wish to acquire to our own confederacy any one or
more of the Spanish provinces? I candidly confess,
that I have ever looked on Cuba as the most interesting addition which could ever be made to our system
of States. The control which, with Florida Point, this
island would give us over the Gulf of Mexico, and the
countries and isthmus bordering on it, as well as all
those whose waters flow into it, would fill up the meas-

ure of our political well being. Yet, as I am sensible
that this can never be obtained, even with her own
consent, but by war ; and its independence, which is
our second interest, (and especially its independence of
England,) can be secured without it, I have no hesitation in abandoning my first wish to future chances, and
accepting its independence, with peace and the friendship of England, rather than its association, at the
expense of war and her enmity.

I could honestly, therefore, join in the declaration
proposed, that we aim not at the acquisition of any of
those possessions ; that we will not stand in the way of
any amicable arrangement between them and the mother
country ; but that we will oppose, with all our means,
the forcible interposition of any other power, as auxil-
iary, stipendiary, or under any other form or pretext ;
and most especially, their transfer to any power by
conquest, cession, or acquisition in any other way. I
should think it, therefore, advisable that the Executive
should encourage the British Government to a continuance in the dispositions expressed in these letters, by
an assurance of his concurrence with them as far as his
authority goes ; and that as it may lead to war, the
declaration of which requires an act of Congress, the
case shall be laid before them for consideration at their
first meeting, and under the reasonable aspect in which
it is seen by himself.

I have been so long weaned from political subjects,
and have so long ceased to take any interest in them,
that I am sensible I am not qualified to offer opinions on
them worthy of any attention. But the question now
proposed involves consequences so lasting, and effects
so decisive of our future destinies, as to re-kindle all
the interest I have heretofore felt on such occasions ;

and to induce me to the hazard of opinions, which will prove only my wish to contribute still my mite towards any thing which may be useful to our country. And praying you to accept it at only what it is worth, I add the assurance of my constant and affectionate friendship and respect.

Th : Jefferson.

PRESIDENT MONROE'S MESSAGE, DECEMBER 2, 1823

In the discussions . . . the occasion has been judged proper for asserting, as a principle in which the rights and interests of the United States are involved, that the American continents, by the free and independent condition which they have assumed and maintain, are henceforth not to be considered as subjects for future colonization by any European powers. . . . The citizens of the United States cherish sentiments the most friendly in favor of the liberty and happiness of their fellow-men on that side of the Atlantic. In the wars of the European powers, in matters relating to themselves, we have never taken any part ; nor does it comport with our policy to do so. It is only when our rights are invaded, or seriously menaced, that we resent injuries, or make preparation for our defense. With the movements in this hemisphere we are of necessity more immediately connected, and by causes which must be obvious to all enlightened and impartial observers. The political system of the allied powers is essentially different in this respect from that of America. This difference proceeds from that which exists in their respective governments. And to the defense of our own, which has been achieved by the loss of so much blood and treasure, and matured by the wisdom

of their most enlightened citizens, and under which we
have enjoyed unexampled felicity, this whole nation is
devoted. We owe it, therefore, to candor, and to the
amicable relations existing between the United States
and those powers, to declare that we should consider
any attempt on their part to extend their system to any
portion of this hemisphere as dangerous to our peace
and safety. With the existing colonies or dependencies of any European power we have not interfered, and
shall not interfere. But with the governments who
have declared their independence, and maintained it,
and whose independence we have, on great consideration and on just principles, acknowledged, we could not
view any interposition for the purpose of oppressing
them, or controlling in any other manner their destiny,
by any European power, in any other light than as the
manifestation of an unfriendly disposition toward the
United States. . . . Our policy in regard to Europe, which was adopted at an early stage of the wars
which have so long agitated that quarter of the globe,
nevertheless remains the same, which is, not to interfere with the internal concerns of any of its powers ; to
consider the government de facto, as the legitimate
government for us ; to cultivate friendly relations with
it, and to preserve those relations by a frank, firm, and
manly policy, meeting in all instances, the just claims
of every power, submitting to injuries from none. But
in regard to these continents, circumstances are eminently and conspicuously different. It is impossible
that the allied powers should extend their political
system to any portion of either continent without en-
dangering our peace and happiness ; nor can any one
believe that our southern brethren, if left to themselves,
would adopt it of their own accord. It is equally im-

possible, therefore, that we should behold such interposition in any form, with indifference.

NULLIFICATION ORDINANCE OF SOUTH CAROLINA, NOV. 24, 1832.

An Ordinance to Nullify certain acts of the Congress
of the United States, purporting to be laws laying
duties and imposts on the importation of foreign commodities.

Whereas the Congress of the United States, by
various acts, purporting to be acts laying duties and
imposts on foreign imports, but in reality intended for
the protection of domestic manufactures, and the giving
of the bounties to classes and individuals engaged in
particular employments, at the expense and to the injury and oppression of other classes and individuals,
and by wholly exempting from taxation certain foreign
commodities, such as are not produced or manufactured in the United States, to afford a pretext for imposing higher and excessive duties on articles similar
to those intended to be protected, hath exceeded its just
powers under the Constitution, which confers on it no
authority to afford such protection ; and hath violated
the true meaning and intent of the Constitution, which
provides for equality in imposing the burthens of taxation upon the several States and portion of the confederacy ; And whereas, the said Congress, exceeding
its just power to impose taxes and collect revenue, for
the purpose of effecting and accomplishing the specific
objects and purposes, which the Constitution of the
United States authorizes it to effect and accomplish,
hath raised and collected unnecessary revenue for objects unauthorized by the Constitution:

We therefore, the people of the State of South Carolina in Convention assembled, do declare and ordain,
and it is hereby declared and ordained, that the several
acts and parts of acts of the Congress of the United
States, purporting to be laws for the imposing of duties
and imposts on the importation of foreign commodities,
and now having actual operation and effect within the
United States, and, more especially, an act entitled
" An act in alteration of the several acts imposing duties on imports," approved on the nineteenth day of
May, one thousand eight hundred and twenty-eight,
and also an act entitled " An act to amend the several
acts imposing duties on imports," approved on the
fourteenth day of July, one thousand eight hundred and
thirty-two, are unauthorized by the Constitution of the
United States, and violate the true meaning and intent
thereof, and are null, void, and no law, nor binding
upon this State, its officers, or citizens ; and all promises, contracts, and obligations, made or entered into,
or to be made or entered into, with purpose to secure
the duties imposed by the said acts, and all judicial
proceedings which shall be hereafter had in affirmance
thereof, are and shall be held utterly null and void.

And it is further ordained, that it shall not be lawful
for any of the constituted authorities, whether of this
State or of the United States, to enforce the payment
of duties imposed by the said acts within the limits of
this State ; but it shall be the duty of the Legislature
to adopt such measures and pass such acts as may be
necessary to give full effect to this ordinance, and to
prevent the enforcement, and arrest the operation of
the said acts and parts of acts of the Congress of the
United States within the limits of this State, from and
after the 1st day of February next; and the duty of all

other constituted authorities, and of all persons residing or being within the limits of this State, and they are hereby required and enjoined, to obey and give effect to this ordinance, and such acts and measures of the Legislature as may be passed or adopted in obedience thereto.

And it is further ordained, that in no case of law or equity decided in the courts of this State, wherein shall be drawn in question the authority of this ordinance, or the validity of such act or acts of the Legislature as may be passed for the purpose of giving validity thereto, or the validity of the aforesaid acts of Congress, imposing duties, shall any appeal be taken or allowed to the Supreme Court of the United States, nor shall any copy of the record be permitted or allowed for that purpose ; and if any such appeal shall be attempted to be taken, the courts of this State shall proceed to execute and enforce their judgments, according to the laws and usages of the State, without reference to such attempted appeal, and the person or persons attempting to take such appeal may be dealt with as for a contempt of the court.

And it is further ordained, that all persons now holding office of honor, profit, or trust, civil or military, under this State, (members of the Legislature excepted,) shall, within such time, and in such manner as the Legislature shall prescribe, take an oath well and truly to obey, execute, and enforce, this ordinance, and such act or acts of the Legislature as may be passed in pursuance thereof, according to the true intent and meaning of the same ; and on the neglect or omission of any such person or persons so to do, his or their office or offices shall be forthwith vacated, and shall be filled up as if such person or persons were dead or had resigned;

and no person hereafter elected to any office of honor, profit, or trust, civil or military, (members of the Legislature excepted,) shall, until the Legislature shall otherwise provide and direct, enter on the execution of his office, or be in any respect competent to discharge the duties thereof, until he shall, in like manner, have taken a similar oath ; and no juror shall be empanneled in any of the courts of this State, in any cause in which shall be in question this ordinance, or any act of the Legislature passed in pursuance thereof, unless he shall first, in addition to the usual oath, have taken an oath that he will well and truly obey, execute, and enforce this ordinance, and such act or acts of the Legislature as may be passed to carry the same into operation and effect, according to the true intent and meaning thereof.

And we, the people of South Carolina, to the end that it may be fully understood by the Government of the United States, and the people of the co-States, that we are determined to maintain this, our ordinance and declaration, at every hazard, do further declare that we will not submit to the application of force, on the part of the Federal Government, to reduce this State to obedience ; but that we shall consider the passage by Congress, of any act authorizing the employment of a military or naval force against the State of South Carolina, her constituted authorities or citizens ; or any act abolishing or closing the ports of this State, or any of them, or otherwise obstructing the free ingress and egress of vessels to and from the said ports, or any other act on the part of the Federal Government, to coerce the State, shut up her ports, destroy or harass her commerce, or to enforce the acts hereby declared to be null and void, otherwise than through the civil tribunals of

the country, as inconsistent with longer continuance of
South Carolina in the Union ; and that the people of
this State will thenceforth hold themselves absolved
from all further obligation to maintain or preserve their
political connexion with the people of the other States,
and will forthwith proceed to organize a separate government, and do all other acts and things which sovereign and independent States may of right do.

Done in convention at Columbia, the twenty-fourth
day of November, in the year of our Lord one thousand
eight hundred and thirty-two, and in the fifty-seventh
year of the declaration of the independence of the
United States of America.

DIGEST OF THE ADDRESS OF THE SOUTH CAROLINA CONVENTION TO THE PEOPLE OF THAT STATE.

The Constitution of the United States, as is admitted
by contemporaneous writers, is a compact of sovereign
States. Though the subject matter of that compact
was a Government, the powers of which Government
were to operate to a certain extent upon the people of
those sovereign States aggregately, and not upon the
State authorities, as is usual in confederacies, still the
Constitution is a confederacy. First. It is a confederacy, because, in its foundations, it possesses not one
feature of nationality. The people of the separate
States, as distinct political communities, ratified the
Constitution, each State acting for itself, and binding
its own citizens, and not those of any other State. The
act of ratification declares it " to be binding on the
States so ratifying." The States are its authors ; their
power created it ; their voice clothed it with authority;
the Government it formed is in reality their govern-

ment, and the union of which it is the bond, is a union
of States not of Individuals.   Secondly.   It is a confederacy, because the extent of the powers of the Government depends, not upon the people of the United States
collectively, but upon the State legislatures, or on the
people of the separate States acting in their State Conventions, each State being represented by a single vote.

The only foundation upon which we can safely erect
the right of a State to protect its citizens, is, that South
Carolina, by the declaration of independence, became,
and has since continued, a free, sovereign, and independent State.   That, as a sovereign State, she has
inherent power to do all those acts which, by law of
nations, any prince or potentate may of right do.
That, like all independent States, she neither has, or
ought she to suffer any other restraint upon her sovereign will and pleasure, than those high moral obligations under which all princes and States ˙ are bound
before God and man, to perform their solemn pledges.
The inevitable conclusion from what has been said,
therefore, is, that, as in all cases of compact between
independent sovereigns, where, from the very nature
of things, there can be no common judge or umpire,
each sovereign has a right " to judge as well of infractions, as of the mode and measure of redress ; " so in
the present controversy between South Carolina and
the Federal Government it belongs solely to her, by
her delegates in solemn convention assembled, to decide whether the Federal compact be violated, and what
remedy the State ought to pursue.   South Carolina,
therefore, cannot and will not yield to any department
of the Federal Government, and still less to the Supreme
Court of the United States, the creature of a Government, which itself is a creature of the States—a right

which enters into the essence of all sovereignty, and
without which it would become a bauble and a name.

It is fortunate for the view we have just taken, that
the history of the Constitution, as traced through the
journals of the Convention which framed the instrument, places the right contended for upon the same
sure foundation.   These journals furnish abundant
proof that no line of jurisdiction between the States and
the Federal Government, " in doubtful cases " could
be agreed upon.   It was conceded by Mr. Madison and
Mr. Randolph, the most prominent advocates for a Supreme Government, that it was impossible to draw this
line, because no tribunal sufficiently impartial, as they
conceived, could be found, and that there was no
alternative but to make the Federal Government supreme, by giving it in all such cases, a negative on all
the acts of the State Legislatures.   The pertinacity
with which this negative power was insisted on by the
advocates of a national government, even after all the
important provisions of the Judiciary or third article of
the Constitution were arranged and agreed to, proves,
beyond doubt, that the Supreme Court was never contemplated by either party in that Convention as an
arbiter to decide conflicting claims of sovereignty between the States and Congress ; and the repeated rejection to take from the States the power of placing their
own construction upon the articles of union, evinces that
the States were resolved never to part with the right to
judge whether the acts of the Federal Legislature were
or were not an infringement of those articles.

Correspondent with the right of a sovereign State to
judge of the infractions of the Federal contract, is the
duty of this Convention to declare the extent of the
grievance, and the mode and measure of redress.   On

both these points, public opinion has already anticipated us in much that we could urge. It is doubted
whether, in any country, any subject has undergone,
before the people, a more thorough examination than
the constitutionality of the several acts of Congress for
the protection of domestic manufactures.

It is a striking characteristic in the operation of a
simple and consolidated government, that it protects
manufactures, agriculture, or any other branch of the
public industry ; that it can establish corporations, or
make roads and canals, and patronize learning and arts.
But it would be difficult to show that such was the government which the sages of the Convention designed
for the States. All these powers were proposed to be
given to Congress, and they were proposed by that
party in the Convention who desired a firm national
government. The Convention having decided on the
Federal form, in exclusion of the national, all these propositions were rejected ; and yet we have lived to see an
American Congress, who can hold no power except by
express grant, as fully in the exercise of these powers
as if they were part and parcel of their expressly delegated authority. Under a pretence of regulating commerce, they would virtually prohibit it. Were this
regulation of commerce resorted to, as a means of coerc-
ing foreign nations to a fair reciprocity in their intercourse with us, or for some other bona fide commercial
purpose, as has been justly said by our Legislature, the
tariff acts would be constitutional. But none of these
acts have been passed as countervailing or retaliatory
measures for restrictions placed on our commerce by
foreign nations. Whilst other nations seem disposed
to relax in their restraints on trade, our Congress seems
absolutely bent upon the interdiction of those articles

of merchandise which were exchangeable for products
of Southern labor ; thus causing the principal burthen
of taxation to fall upon this portion of the union ; and
by depriving us of our accustomed markets to impoverish our whole Southern country.

It is in vain to imagine that with a people who have
struggled for freedom, and know its inestimable value,
such a state of affairs can be endured longer than there
is a well founded hope that reason and justice will resume their empire in the common counsel of the Confederacy.   That hope having expired with the last
session of Congress, by the present tariff act distinctly
and fully recognising, as the permanent policy of the
country, the odious principle of protection, it occurs to
us that there is but one course for the State to pursue.
That course, fellow citizens, is resistance ; not physical,
but moral resistance ; not resistance in an angry or irritated feeling, but resistance by counter legislation ;
which, while it shall evince to the world that our measures are built upon the necessity of tendering to Congress
an amicable issue, to try a doubtful question between
friends and neighbors, shall at the same time secure us
in the enjoyment of our rights and privileges.   It matters not, fellow citizens, by what name this counter
legislation shall be designated—call it nullification,
State interposition, State veto, or by what other name
you please, still, if it be but a resistance to an oppressive measure, it is the course which duty, patriotism
and self-preservation prescribe.   If we are asked upon
what grounds we place the right to resist a particular
law of Congress, and yet regard ourselves as a constituent member of the Union, we answer on the ground
of the compact.   We do not choose in a case of this
kind to recur to what are called our natural rights or

the right of revolution.   We claim to nullify by a more imposing title.   We claim it as a Constitutional right ; not meaning that we derive the right from the Constitution; but we claim to exercise it as one of the parties to the compact ; as not inconsistent with its letter, its genius, and its spirit ; it being distinctly understood at the time of ratifying the Constitution, that the exercise of all sovereign rights not agreed to be had conjointly, was to be exerted separately by the States.

DIGEST OF THE APPEAL TO THE PEOPLE OF THE UNITED STATES.

South Carolina produces almost exclusively agricultural staples which derive their principal value from the demand for them in foreign countries.   Under these circumstances her natural markets are abroad ; and restrictive duties imposed upon her intercourse with those markets, diminish the exchangeable value of her productions very nearly to the full extent of those duties.

Under a system of free trade, the aggregate crop of South Carolina could be exchanged for a larger quantity of manufactures, by at least one third, than it can be now exchanged for under the protecting system.   It is no less evident that the value of that crop is diminished by the protecting system, very nearly, if not precisely, to the extent that the aggregate quantity of manufactures, which can be obtained for it, is diminished.   It is indeed strictly and philosophically true that the quantity of consumable commodities which can be obtained for the cotton and rice, annually produced by the industry of the State, is the precise measure of their aggregate value.   But for the prevalent and habitual error of confounding the money price with the exchangeable value of agricultural staples, these propositions would be regarded as self-evident. If the protecting duties were repealed, one hundred bales of cotton or one hundred barrels of rice would purchase as large a quantity of manufactures as one hundred and fifty would now purchase. The annual income of the State, its means of purchasing and consuming the necessaries and comforts, and luxuries of life, would be increased in a corresponding degree.

Almost the entire cotton crop of South Carolina, amounting annually to more than six millions of dollars, is ultimately exchanged either for foreign manufactures, subject to protecting duties, or for similar domestic manufactures. The natural value of that crop would be all the manufactures which we could obtain for it under a system of unrestricted commerce. The artificial value, produced by the unjust and unconstitutional legislation of Congress, is only such part of those manufactures as will remain after paying a duty of fifty per cent. to the Government ; or to speak with more precision, to the Northern manufacturers. To make this obvious, let it be supposed that the whole of the present crop should be exchanged by the planters themselves for those foreign manufactures, for which it is destined, by the inevitable course of trade, to be ultimately exchanged, either by themselves or by their agents. Let it also be assumed, in conformity with the facts of the case, that New Jersey produces of the very same description of manufactures, a quantity equal to that which is purchased by the cotton crop of South Carolina. We have then two States, of the same Confederacy, bound to bear an equal share of the burthen, and entitled to enjoy an equal share of the benefits of the common government, with precisely the same quantity of productions, of the same quality and kind, produced by their lawful industry. We appeal to your candor, and to your sense of justice, to say whether South Carolina has not a title as sacred and indefeasible to the full and undiminished enjoyment of these productions of her industry, acquired by the combined operations of agriculture and commerce, as New Jersey can have to the like enjoyment of similar productions of her industry, acquired by the process of manufacture. Upon no principle of Constitutional right, upon no principle of human reason or justice, can any discrimination be drawn between the titles of South Carolina and New Jersey to these productions of their capital and labor. Yet what is the discrimination actually made by the unjust, unconstitutional and partial legislation of Congress? A duty on an average of fifty per cent. is imposed upon the productions of South Carolina, while no duty at all is imposed upon the similar productions of New Jersey. The inevitable result is that the manufactures thus lawfully acquired by the honest industry of South Carolina, are worth annually three millions of dollars less to her citizens than the very same quantity, of the very same description of manufactures, are worth to the citizens of New Jersey ; a difference of value produced exclusively by the operation of the protecting system.

Even the States most deeply interested in the maintenance of the protective system will admit, that it is the interest of South Carolina to carry on a commerce of exchanges with foreign countries, free from restrictions, prohibitory burthens, or incumbrances of any kind. We feel and we know that the vital interests of the State are involved in such commerce. It would be a downright insult to our understandings to tell us that

20

our interests are not injured, deeply injured, by these
prohibitory duties, intended and calculated to prevent
us from obtaining the cheap manufactures of foreign
countries for our staples, and to compel us to receive
for them the dear manufactures of our domestic establishments ; or pay the penalty of the protecting duties
for daring to exercise one of the most sacred of our
natural rights. What right, then, human or divine,
have the manufacturing States—for we regard the
Federal Government as a mere instrument in their
hands—to prohibit South Carolina, directly or indirectly, from going to her natural markets, and exchanging the rich productions of her soil, without restriction
or encumbrance, for such foreign articles as will most
conduce to the wealth and prosperity of her citizens ?
It will not, surely, be pretended—for truth and decency
equally forbid the allegation—that, in exchanging our
productions for the cheaper productions of Europe, we
violate any right of the domestic manufacturers, however gratifying it might be to them, if we would purchase their inferior productions at higher prices.

Upon what principle, then, can the State of South
Carolina be called upon to submit to a system which
excludes her from her natural markets, and the mani-
fold benefits of that enriching commerce which a kind
and beneficent Providence has provided to connect her
with the family of nations, by the bonds of mutual interests ?   But one answer can be given to this question.

We implore you, and particularly the manufacturing
States, not to believe that we have been actuated, in
adopting this resolution, by any feeling of resentment
or hostility towards them, or by a desire to dissolve the
political bonds which have so long united our common
destinies. We still cherish that rational devotion to

the Union by which this State has been pre-eminently distinguished in all times past. But that blind and idolatrous devotion which would bow down and worship oppression and tyranny, veiled under a consecrated title, if it ever existed among us, has now vanished forever. Constitutional liberty is the only idol of our political devotion.

We believe that, upon every just and equitable principle of taxation, the whole list of protected articles should be imported free of all duty ; and that the revenue derived from import duties should be raised exclusively from the unprotected articles, or that whenever a duty is imposed upon protected articles imported, an excise duty of the same rate should be imposed upon all similar articles manufactured in the United States. This would be as near an approach to perfect equality as could be possibly made in a system of indirect taxation. No substantial reason can be given for subjecting manufactures obtained from abroad, in exchange for the productions of South Carolina, to the smallest duty, even for revenue, which would not show that similar manufactures made in the United States, should be subject to the very same rate of duty. The former, not less than the latter, are, to every rational intent, the productions of domestic industry, and the mode of acquiring the one is as lawful, and more conducive to the public prosperity than that of acquiring the other.

But we are willing to make a large offering to preserve the Union ; and, with a distinct declaration that it is a concession on our part, we will consent that the same rate of duty may be imposed upon the protected articles that shall be imposed upon the unprotected, provided that no more revenue be raised than is necessary to meet the demands of the government for constitutional purposes, and provided, also, that a duty, substantially uniform, be imposed upon all foreign imports. It is obvious that even under this arrangement, the manufacturing States would have a decided advantage over the planting States ; for it is demonstrably evident that, as communities, the manufacturing States would bear no part of the burthens of federal taxation, so far as the revenue should be derived from protected articles. The earnestness with which their representatives seek to increase the duties on those articles, is conclusive proof that those duties are bounties, and not burthens, to their constituents. As at least two-thirds of the Federal revenue would be raised from protected articles, under the proposed modification of the tariff, the manufacturing States would be entirely exempted from all participation in that proportion of the public burthens.

JACKSON'S PROCLAMATION.

(Written by Edward Livingston.)

I, Andrew Jackson, President of the United States, have thought proper to issue this my Proclamation, stating my views of the Constitution and laws applicable to the measures adopted by the Convention of South Carolina, and to the reasons they have put forth to sustain them, declaring the course which duty will require me to pursue, and, appealing to the understanding and patriotism of the people, warn them of the consequences that must inevitably result from an observance of the dictates of the Convention.

The Ordinance is founded, not on the indefeasible right of resisting acts which are plainly unconstitutional, and too oppressive to be endured ; but on the strange position that any one may not only declare an act of Congress void, but prohibit its execution—that they may do this consistently with the Constitution— that the true construction of that instrument permits a State to retain its place in the Union, and yet be bound by no other of its laws than those it may choose to consider as constitutional. It is true, they add, that to justify this abrogation of a law, it must be palpably contrary to the Constitution ; but it is evident, that, to give the right of resisting laws of that description, coupled with the uncontrolled right to decide what laws deserve that character, is to give the power of resisting all laws. For, as by the theory, there is no appeal, the reasons alleged by the State, good or bad, must prevail. If it should be said that public opinion is a sufficient check against the abuse of this power, it may be asked why it is not deemed a sufficient guard against the passage of an unconstitutional act of Congress ? There is, however, a restraint in this last case, which makes the assumed power of a State more indefensible, and which does not exist in the other. There are two appeals from an unconstitutional act passed by Congress—one to the Judiciary, the other to the people and the States. There is no appeal from the State decision in theory, and the practical illustration shows that the courts are closed against an application to review it ; both judges and jurors being sworn to decide in its favor. But reasoning on this subject is superfluous, when our social compact, in express terms, declares that the laws of the United States, its Constitution, and treaties made under it, are the supreme law of the land ; and, for greater caution, adds " that the judges in every State shall be bound thereby, anything in the Constitution or laws of any State to the contrary notwithstanding.'' And it may be asserted, without fear of refutation, that no Federative Government could exist without a similar provision. . . .

I consider the power to annul a law of the United States, assumed by one State, incompatible with the existence of the Union, contradicted expressly by the letter of the Constitution, unauthorized by its spirit, inconsistent with every principle on which it was founded, and destructive of the great object for which it was formed. . . .

The right to secede is deduced from the nature of the Constitution, which, they say, is a compact between sovereign States, who have preserved their whole sovereignty, and, therefore, are subject to no superior ; that, because they made the compact, they can break it when, in their opinion, it has been departed from by the other States. Fallacious as this course of reasoning is, it enlists State pride, and finds advocates in the honest prejudices of those who have not studied the nature of our Government sufficiently to see the radical error on which it rests.

The people of the United States formed the Constitution, acting through the State legislatures in making the compact, to meet and discuss its provisions, and acting in separate Conventions when they ratified those provisions ; but the terms used in its construction show it to be a government in which the people of all the States collectively are represented. We are one people in the choice of the President and Vice-President. Here the States have no other agency than to direct the mode in which the votes shall be given. The candidates having a majority of all the votes are chosen. The electors of a majority of States may have given their votes for one candidate, and yet another may be chosen. The people, then, and not the States, are represented in the Executive branch.

In the House of Representatives there is this difference, that the people of one State do not, as in the case of President and Vice-President, all vote for the same officers. The people of all the States do not vote for all the members, each State electing only its own representatives. But this creates no material distinction. When chosen, they are all representatives of the United States, not representatives of the particular State from which they come. They are paid by the United States, not by the State, nor are they accountable to it for any act done in the performance of their legislative functions ; and however they may in practice, as it is their duty to do, consult and prefer the interests of their particular constituents when they come in conflict with any other partial or local interest, yet it is their first and highest duty, as representatives of the United States, to promote the general good.

The Constitution of the United States then forms a government, not a league ; and whether it be formed by compact between the States, or in any other manner, its character is the same. It is a government in which all the people are represented, which operates directly on the people individually ; not upon the States—they retained all the power they did not grant. But each State having expressly parted with so many powers as to constitute, jointly with the other States, a single nation, cannot, from that period, possess any right to secede, because such secession does not break a league, but destroys the unity of a nation ; and any injury to that unity is not only a breach which would result from the contravention of a compact ; but it is an offense against the whole Union. To say that any

State may at pleasure secede from the Union, is to say that the United States are not a nation ; because it would be a solecism to contend that any part of a nation might dissolve its connection with the other parts, to their injury or ruin, without committing any offence. Secession, like any other revolutionary act, may be morally justified by the extremity of oppression ; but to call it a constitutional right, is confounding the meaning of terms ; and can only be done through gross error, or to deceive those who are willing to assert a right, but would pause before they made a revolution, or incurred the penalties consequent on a failure.

Because the Union was formed by compact, it is said the parties to that compact may, when they feel themselves aggrieved, depart from it ; but it is precisely because it is a compact that they cannot. A compact is an agreement or binding obligation. It may by its terms have a sanction or penalty for its breach, or it may not. If it contains no sanction, it may be broken with no other consequence than moral guilt : if it have a sanction, then the breach insures the designated or implied penalty. A league between independent nations, generally, has no sanction other than a moral one ; or if it should contain a penalty, as there is no common superior, it cannot be enforced. A government on the contrary, always has a sanction, express or implied ; and, in our case, it is both necessarily implied and expressly given. An attempt, by force of arms, to destroy a government, is an offence, by whatever means the constitutional compact may have been formed, and such government has the right, by the law of self-defence, to pass acts for punishing the offender, unless the right is modified, restrained, or resumed by the constitutional act. In our system,

although it is modified in the case of treason, yet
authority is expressly given to pass all laws necessary
to carry its powers into effect, and, under this grant,
provision has been made for punishing acts which ob-
struct the due administration of the laws. It would
seem superfluous to add anything to show the nature
of that union which connects us ; but, as erroneous
opinions on this subject are the foundation of doctrines
the most destructive to our peace, I must give some
further development to my views on this subject. No
one, fellow citizens, has a higher reverence for the reserved rights of the States than the magistrate who
now addresses you. No one would make any greater
personal sacrifices, or official exertions, to defend them
from violation ; but equal care must be taken to prevent, on their part, an improper interference with, or
resumption of, the rights they have vested in the nation. The line has not been so distinctly drawn as to
avoid doubts in some cases of the exercise of power.
Men of the best intentions and soundest views may
differ in their construction of some parts of the Constitution ; but there are others on which dispassionate reflection can leave no doubt. Of this nature appears to
be the assumed right of secession. It rests, as we have
seen, on the alleged undivided sovereignty of the
States, and on their having formed, in this sovereign
capacity, a compact which is called the Constitution,
from which, because they made it, they have the
right to secede. Both of these positions are erroneous,
and some of the arguments to prove them so have been
anticipated.

The States severally have not retained their entire
sovereignty. It has been shown that, in becoming
parts of a nation, not members of a league, they sur-

rendered many of their essential parts of sovereignty.
The right to make treaties, declare war, levy taxes,
exercise exclusive judicial and legislative powers, were
all of them functions of sovereign power. The States,
then, for all these purposes were no longer sovereign.
The allegiance of their citizens was transferred, in the
first instance, to the government of the United States ;
they became American citizens, and owed obedience to
the Constitution of the United States, and to laws made
in conformity with the powers it vested in Congress.
This last position has not been, and cannot be denied.
How, then, can that State be said to be sovereign and
independent whose citizens owe obedience to laws not
made by it, and whose magistrates are sworn to disregard those laws when they come in conflict with those
passed by another ?   What shows conclusively that the
States cannot be said to have reserved an undivided
sovereignty, is, that they expressly ceded the right to
punish treason, not treason against their separate
power, but treason against the United States.   Treason
is an offense against sovereignty, and sovereignty must
reside with the power to punish it.   But the reserved
rights of the States are not less sacred because they
have, for their common interest, made the general government the depository of these powers.

The unity of our political character (as has been
shown for another purpose) commenced with its very
existence.   Under the royal government we had no
separate character ; our opposition to its oppressions
began as United Colonies.   We were the United States
under the Confederation, and the name was perpetuated,
and the Union rendered more perfect, by the Federal
Constitution.   In none of these stages did we consider
ourselves in any other light than as forming one nation.

Treaties and alliances were made in the name of all.
Troops were raised for the joint defence. How, then,
with all these proofs, that under all changes of our position we had, for designated purposes and defined
powers, created national governments—how is it, that
the most perfect of those several modes of union should
now be considered as a mere league that may be dissolved at pleasure ? It is from an abuse of terms.
Compact is used as synonymous with league, although
the true term is not employed, because it would at once
show the fallacy of the reasoning. It would not do to
say that our Constitution was only a league, but it is
labored to prove it a compact, (which in one sense it
is,) and then to argue that as a league is a compact,
every compact between nations must of course be a
league, and that from such an engagement every sovereign power has a right to recede. But it has been
shown that, in this sense, the States are not sovereign ;
and that even if they were, and the national Constitution had been formed by compact, there would be no
right in any one State to exonerate itself from its obligations.

So obvious are the reasons which forbid this secession, that it is necessary only to allude to them. The
union was formed for the benefit of all. It was produced by mutual sacrifices of interests and opinions.
Can those sacrifices be recalled ? Can the States, who
magnanimously surrendered their title to the territories
of the west, recall the grant ? Will the inhabitants of
the inland States agree to pay the duties that may be
imposed without their assent by those on the Atlantic
or the Gulf, for their own benefit ? Shall there be a
free port in one State, and onerous duties in another ?
No one believes that any right exists in a single State

to involve all the others in these and countless other
evils, contrary to the engagement solemnly made.
Every one must see that the other States, in self-
defence, must oppose it at all hazards.

These are the alternatives that are presented by the
Convention: a repeal of all the acts for raising revenue,
leaving the government without the means of support ;
or an acquiescence in the dissolution of our Union by
the secession of one of its members. When the first
was proposed, it was known that it could not be listened
to for a moment. It was known, if force was applied
to oppose the execution of the laws that it must be repelled by force ; that Congress could not, without involving itself in disgrace and the country in ruin,
accede to the proposition : and yet if this is done in
a given day, or if any attempt is made to execute the
laws, the State is, by the ordinance, declared to be out
of the Union. The majority of a Convention assembled
for the purpose, have dictated these terms, or rather
this rejection of all terms, in the name of the people
of South Carolina. It is true that the governor of the
State speaks of the submission of their grievances to a
Convention of all the States, which, he says they
" sincerely and anxiously seek and desire." Yet this
obvious and constitutional mode of obtaining the sense
of the other States on the construction of the federal
compact, and amending it, if necessary, has never been
attempted by those who have urged the State on to this
destructive measure. The State might have proposed
the call for a general convention to the other States ;
and Congress, if a sufficient number of them concurred,
must have called it. But the first magistrate of South
Carolina, when he expressed a hope that, " on a review
by Congress and the functionaries of the general gov-

ernment of the merits of the controversies," such a
convention will be accorded to them, must have known
that neither Congress, nor any functionary of the general government, has authority to call such a convention, unless it be demanded by two-thirds of the States.
This suggestion, then, is another instance of the reckless inattention to the provisions of the Constitution
with which this crisis has been madly hurried on ; or
of the attempt to persuade the people that a constitutional remedy has been sought and refused. If the
legislature of South Carolina " anxiously desire " a
general convention to consider their complaints, why
have they not made application for it in the way the
Constitution points out ?   The assertion that they
" earnestly seek it," is completely negatived by the
omission.

This then is the position in which we stand. A
small majority of the citizens of one State in the Union
have elected delegates to a State convention ; that convention has ordained that all the revenue laws of the
United States must be repealed, or that they are no
longer a member of the Union. The governor of that
State has recommended to the legislature the raising
of an army to carry the secession into effect, and that
he may be empowered to give clearances to vessels in
the name of the State.   No act of violent opposition to
the laws has yet been committed, but such a state of
things is hourly apprehended ; and it is the intent of
this instrument to proclaim, not only that the duty imposed on me by the Constitution " to take care that the
laws be faithfully executed," shall be performed to the
extent of the powers already invested in me by law, or
of such others as the wisdom of Congress shall devise
and entrust to me for that purpose, but to warn the

citizens of South Carolina who have been deluded into
an opposition to the laws, of the danger they will incur
by obedience to the illegal and disorganizing ordinance
of the convention ; to exhort those who have refused
to support it to persevere in their determination to uphold the Constitution and laws of their country ; and
to point out to all, the perilous situation into which the
good people of that State have been led, and that the
course they are urged to pursue is one of ruin and disgrace to the very State whose rights they affect to
support.

Fellow citizens of my native State, let me not only
admonish you, as the First Magistrate of our common
country, not to incur the penalty of its laws, but use
the influence that a father would over his children
whom he saw rushing to certain ruin. In that paternal language, with that paternal feeling, let me tell
you, my countrymen, that you are deluded by men
who are either deceived themselves, or wish to deceive
you. Mark under what pretences you have been led
on to the brink of insurrection and treason, on which
you stand ! First, a diminution of the value of your
staple commodity, lowered by over-production in other
quarters, and the consequent diminution in the value
of your lands, were the sole effect of the tariff laws.

The effect of those laws was confessedly injurious,
but the evil was greatly exaggerated by the unfounded
theory you were taught to believe, that its burdens
were in proportion to your exports, not to your consumption of imported articles. Your pride was roused
by the assertion that a submission to those laws was a
state of vassalage, and that resistance to them was
equal, in patriotic merit, to the oppositions our fathers
offered to the oppressive laws of Great Britain. You

were told that this opposition might be peaceably—
might be constitutionally made ;—that you might enjoy
all the advantages of the Union, and bear none of its
burthens. Eloquent appeals to your passions, to your
State pride, to your native courage, to your sense of
real injury, were used to prepare you for the period
when the mask which concealed the hideous features of
disunion, should be taken off. It fell, and you were
made to look with complacency on objects which, not
long since, you would have regarded with horror.
Look back to the acts which have brought you to this
state—look forward to the consequences to which it
must inevitably lead ! Look back to what was first
told you as an inducement to enter into this dangerous
course. The great political truth was repeated to you,
that you had the revolutionary right of resisting all laws
that were palpably unconstitutional and intolerably
oppressive; it was added that the right to nullify a law
rested on the same principle, but that it was a peaceable remedy ! This character which was given to it,
made you receive with too much confidence, the asser-
tions that were made of the unconstitutionality of the
law and its oppressive effects. Mark, my fellow citizens, that, by the admission of your leaders, the unconstitutionality must be palpable, or it will not justify
either resistance or nullification ! What is the meaning of the word palpable, in the sense in which it is
here used ? That which is apparent to every one; that
which no man of ordinary intellect will fail to perceive.
Is the unconstitutionality of these laws of that description ? Let those among your leaders who once approved and advocated the principle of protective duties,
answer the question ; and let them choose whether
they will be considered as incapable, then, of perceiving

that which must have been apparent to every man of common understanding, or as imposing upon your confidence, and endeavoring to mislead you now. In either case, they are unsafe guides in the perilous path they urge you to tread. Ponder well on this circumstance, and you will know how to appreciate the exaggerated language they address to you. They are not champions of liberty emulating the fame of our revolutionary fathers ; nor are you an oppressed people, contending, as they repeat to you, against worse than colonial vassalage.

You are free members of a flourishing and happy Union. There is no settled design to oppress you. You have indeed felt the unequal operation of laws which may have been unwisely, not unconstitutionally passed ; but that inequality must necessarily be removed. At the very moment when you were madly urged on to the unfortunate course you have begun, a change in public opinion had commenced. The nearly approaching payment of the public debt, and the consequent necessity of a diminution of duties, had already produced a considerable reduction, and that, too, on some articles of general consumption in your State. The importance of this change was underrated, and you were authoritatively told that no further alleviation of your burthens was to be expected, at the very time when the condition of the country imperiously demanded such a modification of the duties as should reduce them to a just and equitable scale. But, as if apprehensive of the effect of this change in allaying your discontents, you were precipitated into a fearful state in which you now find yourselves.

I have urged you to look back to the means that were used to hurry you on to the position you have now assumed, and forward to the consequences it will produce. Something more is necessary. Contemplate the condition of that country of which you still form an important part. Consider its government, uniting in one bond of common interest and general protection so many different States—giving to all their inhabitants the proud title of American citizens, protecting their commerce, securing their literature and their arts ; facilitating their intercommunication ; defending their frontiers ; and making their name respected in the remotest parts of the earth. Consider the extent of its territory ; its increasing and happy population ; its advance in arts, which render life agreeable ; and the sciences, which elevate the mind ! See education spreading the lights of religion, morality, and general information into every cottage in this wide extent of our Territories and States ! Behold it as the asylum where the wretched and the oppressed find a refuge and support ! Look on this picture of happiness and honor, and say—We too, are citizens of America !—Carolina is one of these proud States—her arms have defended— her best blood has cemented this happy Union ! And then add, if you can, without horror and remorse, this happy Union we will dissolve ; this picture of peace and prosperity we will deface ; this free intercourse we will interrupt ; these fertile fields we will deluge with blood; the protection of that glorious flag we renounce; the very name of America we discard. And for what, mistaken men—for what do you throw away these inestimable blessings ? for what would you exchange your share in the advantages and honor of the Union ? For the dream of separate independence—a dream interrupted by bloody conflicts with your neighbors, and a vile dependence on a foreign power. If your leaders

21

could succeed in establishing a separation, what would
be your situation ?   Are you united at home—are you
free from the apprehension of civil discord, with all its
fearful consequences ?   Do your neighboring republics,
every day suffering some new revolution, or contending
with some new insurrection—do they excite your envy ?
But the dictates of a high duty oblige me solemnly to
announce that you cannot succeed.   The laws of the
United States must be executed.   I have no discretion-
ary power on the subject—my duty is emphatically
pronounced in the Constitution.   Those who told you
that you might peacably prevent their execution, deceived you—they could not have been deceived themselves.   They know that a forcible opposition alone
could prevent the execution of the laws, and they know
that such opposition must be repelled.   Their object
is disunion ; but be not deceived by names ; disunion,
by armed force, is *treason*.   Are you really ready to
incur its guilt ?   If you are, on the heads of the insti-
gators of the act be the dreadful consequences—on their
heads be the dishonor, but on yours may fall the punishment ; on your unhappy State will inevitably fall
all the evils of the conflict you force upon the government of your country.   It cannot accede to the mad
project of disunion, of which you would be the first
victims—its First Magistrate cannot, if he would, avoid
the performance of his duty :—the consequence must be
fearful for you, distressing to your fellow citizens here,
and to the friends of good government throughout the
world.   Its enemies have beheld our prosperity with a
vexation they could not conceal—it was a standing
refutation of their slavish doctrines, and they will point
to our discord with the triumph of malignant joy.   It
is yet in your power to disappoint them.   There is yet

time to show that the descendants of the Pinckneys, the
Sumters, the Rutledges, and of the thousand other
names which adorn the pages of your revolutionary
history, will not abandon that Union, to support which
so many of them fought, and bled, and died.

I adjure you, as you honor their memory—as you
love the cause of freedom, to which they dedicated
their lives—as you prize the peace of your country, the
lives of its best citizens, and your own fair fame, to retrace your steps.   Snatch from the archives of your
State the disorganizing edict of its Convention—bid its
members to reassemble, and promulgate the decided
expressions of your will to remain in the path which
alone can conduct you to safety, prosperity, and honor.
Tell them that, compared to disunion, all other evils
are light, because that brings with it an accumulation
of all.   Declare that you will never take the field unless the star spangled banner of your country shall float
over you ; that you will not be stigmatized when dead,
and dishonored and scorned while you live, as the
authors of the first attack on the Constitution of your
country.   Its destroyers you cannot be.   You may disturb its peace—you may interrupt the course of its
prosperity—you may cloud its reputation for stability,
but its tranquillity will be restored, its prosperity will
return, and the stain upon its national character will
be transferred, and remain an eternal blot on the memory of those who caused the disorder.   .   .   .

May the Great Ruler of Nations grant that the signal
blessings with which he has favored ours, may not, by
the madness of party or personal, ambition, be disregarded and lost ; and may his wise Providence bring
those who have produced this crisis to see their folly,
before they feel the misery of civil strife ; and inspire a

returning veneration for that Union, which, if we may dare to penetrate his designs, he has chosen as the only means of attaining the high destinies to which we may reasonably aspire.

ANDREW JACKSON.

EDWARD LIVINGSTON,
      Secretary of State.
December 10, 1832.

LIVINGSTON'S VIEWS.

I think that the Constitution is the result of a compact, entered into by the several States ; by which they surrendered a part of their sovereignty to the Union, and vested the part so surrendered in a general government.

That this government is partly popular, acting directly on the citizens of the several States, partly federative, depending for its existence and action on the existence and action of the several States.

That by the institution of this government the States have unequivocally surrendered every constitutional right of impeding or resisting the execution of any decree or judgment of the Supreme Court, in any case of law or equity between persons, or on matters of whom or on which that court has jurisdiction, even if such decree or judgment should in the opinion of the States, be unconstitutional.

That, in cases in which a law of the United States may infringe the constitutional right of a State, but which, in its operation, cannot be brought before the Supreme Court, under the terms of the jurisdiction expressly given to it over particular persons or matters, that court is not created the umpire between a State that may deem itself aggrieved and the general government.

That among the attributes of sovereignty retained by the States is that of watching over the operations of the general government, and protecting its citizens against their unconstitutional abuse ; and that this can be legally done,—

First, in the case of an act in the opinion of the State palpably unconstitutional, but affirmed in the Supreme Court in the legal exercise of its functions,—

By remonstrating against it to Congress ;

By an address to the people in their elective functions to change or instruct their representatives ;

By a similar address to the other States, in which they will have a right to declare that they consider the act as unconstitutional and therefore void ;

By proposing amendments to the Constitution in the manner pointed out by that instrument ;

And, finally, if the act be intolerably oppressive, and they find the general government persevere in enforcing it, by a resort to the natural right which every people have to resist extreme oppression.

Secondly, if the act be one of those few which in their operation cannot be submitted to the Supreme Court, and be one that will in the opinion of the State, justify the risk of a withdrawal from the Union, that this last extreme remedy may at once be resorted to.

That the right of resistance to the operation of an act of Congress, in the extreme cases above alluded to, is not a right derived from the Constitiution, but can be justified only on the supposition that the Constititution has been broken, and the State absolved from its obligation ; and that, whenever resorted to, it must be at the risk of all the penalties attached to an unsuccessful resistance to established authority.

That the alleged right of a State to put a veto on the

execution of a law of the United States, which such
State may declare to be unconstitutional, attended (as,
if it exist, it must be) with a correlative obligation on
the part of the general government to refrain from executing it, and the further alleged obligation on the
part of that government to submit the question to the
States by proposing amendments, are not given by the
Constitution, nor do they grow out of any of the reserved powers.

That the exercise of the powers last mentioned, would
introduce a feature in our government, not expressed
in the Constitution, not implied from any right of sovereignty reserved to the States, not suspected to exist
by the friends or enemies of the Constitution when it
was framed or adopted, not warranted by practice or
contemporaneous exposition, nor implied by the true
construction of the Virginia Resolutions in '98.

That the introduction of this feature in our government would totally change its nature, make it ineffi-
cient, invite to dissension, and end, at no distant period,
in separation ; and that, if it had been proposed in the
form of an explicit provision in the Constitution,
would have been unanimously rejected, both in the
Convention which framed that instrument, and in those
which adopted it.

That the theory of the federal government being the
result of the general will of the people of the United
States in their aggregate capacity, and founded, in no
degree, on compact between the States, would tend to
the most disastrous practical results ; that it would
place three-fourths of the States at the mercy of one-
fourth, and lead inevitably to a consolidated government, and finally to monarchy, if the doctrine were
generally admitted.

WEBSTER'S FOUR PROPOSITIONS.

1. That the Constitution is not a league, confederacy or compact between the people of the several States in their sovereign capacities ; but a government proper, founded on the adoption of the people, and creating direct relations between itself and individuals.

2. That no State authority has power to dissolve these relations ; that nothing can dissolve them but revolution ; and that consequently there can be no such thing as secession without revolution.

3. That there is a supreme law, consisting of the Constitution of the United States, acts of Congress passed in pursuance of it, and treaties ; and that in cases not capaple of assuming the character of a suit in law or equity, Congress must judge of and finally interpret this supreme law, so often as it has occasion to pass acts or legislation ; and in cases capable of assuming, and actually assuming, the character of a suit, the Supreme Court of the United States is the final interpreter.

4. That an attempt of a State to abrogate, annul, or nullify an act of Congress, or to arrest its operation within her limits, on the ground that in her opinion such law is unconstitutional, is a direct usurpation of the just powers of the general government, and of the equal rights of the other States, a plain violation of the Constitution, and a proceeding essentially revolutionary in its character and tendency.

CHAPTER VII.

SECESSION IN 1861.

JEFFERSON'S prophecy proved true. The sectional line having been drawn geographically, sectionalism began to organize as such. Churches fell apart across the barrier. There were the Northern and the Southern Presbyterians, Methodists, and Baptists. The question arose once more, " Is Christ divided ? " Political parties strove in vain to straddle without partiality the dividing ridge. At the North, anti-slavery began to be anti Southern-slavery. The South resented meddling with their domestic affairs. Slavery instead of being universally condemned was approved from the rostrum and the pulpit. It could not have endured or tolerated opposition. No one knows how much suppressed sentiment to the contrary existed. Miss Martineau in 1836 says, " Among the many hundreds of persons in the slave States, with whom I conversed on the subject of slavery, I met with only one who defended the institution altogether. All the rest who vindicated its existence, did so on the ground of the impossibility of doing it away." But that was when there was an exact sectional balance of thirteen slave States to thirteen free States. Governor McDuffie of South Carolina about that time declared slavery was the corner-stone of republican liberties. So hopelessly identified was the South becoming with the peculiar institution that it must adjust to it its taste, its conscience, its religion, its politics. Madison remarked that as the Puritan North grew liberal the chivalrous South was growing bigoted.

It took five slaves to do the work of one freeman. The average yield of grain by slave labor was six bushels, to twenty bushels for free labor. The hive was full of laborers to support a few drones. In 1832 the Richmond *Enquirer* said, " God only knows what it is the part of wise men to do on this appalling subject. Of this we can be sure that the difference— nothing short of frightful—between all that exists on one side of the Potomac and all on the other is owing to that alone. The disease is deep seated ; it is at the heart's core ; it is consuming our vitals. What is to be done ? O God ! we don't know ; but something must be done."

The form that labor takes inevitably colors social relations. At the North, democracy and a sentiment of equality ruled. The conflict of capital and labor had not yet gained any headway. The New England clique that assumed to be the " Best," was absorbed in democracy ; and the aristocratic four-hundreds were as yet only subjects of sarcastic comment. At the polls they had no power beyond that of the hard-handed class ; and in general their social relations were not larger or pleasanter or more cultivating than those of the bulk of society. But in the South, as late as 1850 there were only 347,525 slave holders out of six and a quarter million of whites. This aristocracy depressed not only slaves but all non-slave-holding whites. In the nature of things it made labor dishonorable. It

branded the negro as the natural child of toil. The
white man's obligation was to rank above the toilers.
The poor white had no resort- but ignorant conceit.
He must not even help himself above his present condition.

The contrast between the two sections was more
marked in matters of public and free schools. In Connecticut in 1833 the school fund was $1,929,738, and
the dividend was at the rate of one dollar to each child
of school age in the State—these numbering 83,912.
In New York the number of school children in common
schools was 534,000. In Massachusetts the annual
school-money tax was $310,178 besides $150,000 raised
by contribution and $276,575 paid for tuition. In Ken-
tucky the common-school system was created in 1830,
and one school was in operation in 1835. In South
Carolina the number of scholars in public schools at the
same time was 5361. In Maryland one primary school
had been founded, and about ten academies. No public schools existed in North Carolina, Alabama, Mississippi, Missouri, Louisiana, Tennessee; and no free
schools in Virginia. A fund in some States was accumulating to eventuate in a system thereafter.

The suppression of sentiment at the South also
worked disastrously in the way of creating literature.
Boston was the American Athens; and New England
the mother of authors as well as inventive genius.
Whatever the South did produce was mostly belittled
by a defense of its chief fault, rather than ennobled by
an expression of its virtues. Slavery seemed to be
almost the only topic that could be spoken of; and it
was the one topic that called for severest repression.
In oratory the two sections were more nearly matched.
The South developed a class of brilliant but fiery public

speakers, capable of swaying crowds with marvellous skill. Prentiss, Hayne, Calhoun, McDuffie, Clay met on the rostrum Webster, Sumner, Seward, Douglas, Chase, Hale, and Corwin. The effect of this clash of mighty orators at the Capitol was by no means to widen the breach between the two sections. Congress, although the scene of the forensic flashes of clashing opinions, was after all of great value in bringing the leaders into social relations. Miss Martineau wrote, " Deadly political enemies meet at Washington, and swell and declaim at one another. They find themselves some sunny day lying on the grass under a tree. . . . They have discussed nullification ; and yet rise up cordial friends." Washington was admirably located. It was on neutral grounds near the dividing line of the sections. If its conditions favored either side it was that of the weaker.

Harvard and Yale were also doing much to prevent sectionalism from absolutely severing the Union. The South had to send its boys either to the North or to Europe to get a classical finish. They secured a breadth of apprehension by observation ; and broadened their sympathies not only by travel but by college friendships. We have noted that President Washington foresaw the immense value of educating our best youth from diverse localities in a great university, which he hoped to see founded at Washington. In his will he said, " It has been my ardent wish to see a plan devised on a liberal scale to spread systematic ideas through all parts of this rising empire—thereby to do away with local attachments and State prejudices." For this end Washington left in charge of Congress certain " Potomac shares," which would now amount to four or five millions of dollars. It is not improbable that had such a universitiy been founded the sectional strife would never have reached the extreme of hate and war.

More hopeless was the moral division of the two sections. The North believed it was the champion of humanity. It certainly had the sympathy of European philanthropists. Jefferson Davis avers that no moral considerations were really involved in either the earlier or the later controversies. '' It was,'' he says, '' a struggle for the maintenance or the destruction of that balance of power or equipoise between North and South which was early recognized as a cardinal principle in our Federal system.'' The very opposite was fundamental truth. From first to last the sectional struggle inextricably involved moral issues. Oglethorpe, the founder of Georgia, said, '' Slavery is against the Gospel, as well as the fundamental law of England. As trustees we refuse to make a law permitting such a horrid crime.'' William Penn set his slaves free ; and his followers were the earliest abolitionists. Jefferson, lamenting his failure by a single vote to exclude slavery from the Southwest, wrote, '' The voice of a single individual would have prevented this abominable crime. Heaven will not always be silent ; the friends to the rights of human nature will in the end prevail.'' The original draft of the Declaration of Independence in Jefferson's handwriting, contained these words about King George, '' He has waged cruel war against nature itself, violating its most sacred rights of life and liberty, in the persons of a distant people, who never offended him, captivating and carrying them into slavery in another hemisphere. Determined to keep open a market where men should be bought and sold, he has prostituted his negative, by suppressing every legislative attempt to prohibit or restrain this execrable com-

merce.'' Although this clause was stricken out by the
committee, the accepted draft made the eternal corner-
stone of our national existence the belief that all men
are created with equal rights to life, liberty, and the
pursuit of happiness. Franklin was President of the
first abolition society in America. The first convention
of these societies declared that '' freedom and slavery
cannot long exist together.'' Rhode Island abolished
slavery in the spirit of Roger Williams, as '' a trespass
on universal right.'' Not until the cotton-gin created
an increasing demand for negroes in the lower States
did Virginia give up the effort to emancipate. Slavery
forced its defenders to emphasize liberty as the new
watchword for the white masses ; bondage for the
blacks. It compelled them to denounce with vitriolic
bitterness miscegenation of abolitionism, while their
own negro stock was turning white with amazing
rapidity. Jefferson wrote, '' The whole commerce between master and slave is a perpetual exercise of the
most boisterous passions, the most unremitting despotism on the one part, and degrading submissions on the
other.'' Madison described with contempt the eagerness of the clergy to make '' converts of those who,
having no rights, can have no duties.'' But slavery
had become ingrained with Southern life. A young
man, a church member, in a Northern college in 1849
left his studies in the middle of the course, for no other
reason but that he '' seriously missed the free sexual
intercourse with his father's slave women, to which he
had been accustomed.'' Olmstead quotes a Southern
merchant as begging his brother to send his children
North to school, '' for he might as well educate them in
a brothel as in the way they were.'' Chancellor
Harper argued that unchastity among female slaves

was "not a vice, but only a weakness." He said
nothing of the other parties to this sexualism. 200,000
mulattoes were born in the Southern States between
1850 and 1860. In 1850 the proportion of slaves of
mixed blood was ten per cent.; in 1860 it was twelve
per cent. "We Southern ladies," said a sister of President Madison, "are complimented with the name of
wives, but we are only the mistresses of seraglios."
Madison himself said: "Licentiousness only stops short
of the extinction of the race. Every slave girl is expected to be a mother by the time she is fifteen."

As sectionalism grew, this moral consciousness of the
wrongs of slavery was as manifest at the South as at the
North; but in different ways. Although a likely negro
baby born into a family added one hundred dollars to
the family property, there were many who urged that
slave auctions should be less publicly held. John Randolph, when asked who was the most eloquent person
he had ever heard, replied, "A quadroon ; and her rostrum was the auction block."

The slave trader was shunned as a Pariah in every
Southern State. Lincoln with keen incision said in a
speech in 1854, "You have among you a sneaking
individual of the class of native tyrants known as the
slave dealer. He watches your necessities, and crawls
up to buy your slave at a speculating price. If you
cannot help it you sell to him ; but if you can help it
you drive him from your door. You despise him
utterly. You do not recognize him as a friend, or
even as an honest man. Your children may rollic with
the little negroes, but not with the slave dealer's children. It is common to join hands with men you meet ;
but with him you avoid the ceremony—instinctively
shrinking from the snaky contact. Why is this ? You

do not so treat the man who deals in corn or cattle or tobacco.''

Not only had the South manifestations of conscience on the subject of slavery, but the North developed some phases of sentiment on the subject, that were unmitigated meanness or hypocrisy. While it had abolished human bondage under the law, it bound men by social customs to a slavery as hopeless as that at the South. In only six free States could a free black man cast a vote. He could not enter the colleges on any terms, before Oberlin was founded. In the churches he could have his soul saved by a white preacher, only while sitting in a corner pew set apart for '' niggers.''

The North, desiring to avoid extreme partisanship, mobbed those who desired to meddle with Southern affairs, while the South developed a peculiar breed of compromisers, of which Clay and Crittenden and Bell were conspicuous examples. Slavery, while recognized as sectional, was of such a nature that it could not be localized. Slaves ran away continuously. The Northern border States enacted stringent laws for their rendition, but few were ever caught after they were out of sight of the Ohio River. The underground railroad hastened them from point to point till they reached Canada. When overtaken, fugitives often committed suicide rather than submit to capture and its consequences. There were professional abettors who crossed into the slave States in disguise to run off negroes. By the laws of the Southern States slaves were as much property as horses or cattle. By the laws of the Northern States human beings could not be property. Over both the Constitution unquestionably recognized the Southern principle as correct. Garrison retorted that the Constitution was '' a covenant with Death and a league with Hell." It clearly was not a simple matter to bind together in one Union a double church, double social order, and double forms of labor, creating diverse consciences, diverse political aims, and antagonistic commercial interests.

Anti-slavery agitation broke out with desperate earnestness in 1830–31. It was just at this time that an insurrection of negroes occurred in Virginia and North Carolina. Randolph declared " the fire bell never tolled without sending a thrill of terror through Richmond." Miss Martineau noticed that any sudden alarm at a theatre sent every one home as quickly and quietly as possible. They indulged in no curiosity ; but feared the worst. It was necessary to keep the slaves in abject ignorance lest they find out the road to freedom. It followed that incendiary, that is abolition reading matter, should be kept out of the whole South. In 1835 Mrs. Barbauld's works were returned to Northern publishers, because they contained a dialogue between master and slave. One of Miss Sedgwick's novels was interdicted.

During Jackson's administration the question arose of rifling the United States mails. Kendall, the unscrupulous Postmaster-General, wrote that any postmaster owed not only obedience to the country but to " higher " laws, to protect his community from danger. " Entertaining these views I cannot sanction, and will not condemn the step you have taken." This was practically the first step in the drama, after the Missouri Compromise. Jackson in a message recommended a general law repressing incendiary matter. Calhoun, as chairman of a select committee to whom the matter was referred in 1836, reported that if Congress could prevent the circulation, it could also enforce the circulation of any such matter as it chose. He recommended that each State be considered judge of what was calculated to disturb its security. The matter dropped in Congress; but it did not drop as a practical fact that mails were never afterward sacred in the slave States.

The right to discuss slavery in the pulpit or press next became the great issue. Virginia by Legislative resolution in 1836 asked the Northern States to "adopt penal enactments to suppress abolition societies. The Augusta *Chronicle* of 1833 said, "The Southern States must at once declare that the discussion of slavery will compel them to secede and appeal to the sword." Lovejoy, editor of a religious newspaper in St. Louis, commented on an editorial that appeared in a daily paper, urging the revision of the State Constitution to exclude slavery. His strictures were not severe on slavery, but were logical and courageous. This led on to riots, and finally in 1836 to his assassination. Garrison, who had established *The Liberator* was mobbed in Boston; a rope was put round his neck and he was dragged through the streets. A poor truckman seized a stake and drove off the mob, composed of "gentlemen." The controversy was transferred to Congress by a tireless series of petitions to abolish slavery in the District of Columbia. The right of such petition was denied. "The Crisis" called for a convention of Southern States to enable them to act together. John Quincy Adams headed the party that would not be forbidden free right to petition. In 1842 Joshua R. Giddings stood beside the sage of Massachusetts; but he was formally censured by the House. He resigned; and his constituents sent him back by a vote practically unanimous. This battle

22

of Congress with annoying papers extended from 1830 to 1845.

Meanwhile the Texas war of independence had eventuated in an independent republic. In 1837, while Van Buren was President, this republic proposed annexation to the United States. A negative was returned; but the proposition was not only not objectionable to the South, but was soon the pet object of most of her leaders. Tyler openly advocated annexation. Jackson wrote from retirement in 1843 urging the same step. Clay opposed it as unjust to Mexico. He specifically declared that he could conceive no motive for the acquisition of Texas more unfortunate or fatal than that of strengthening one part against another part of the States. "Such a principle, put into practical operation, would menace the existence, if it did not sow the seeds of a dissolution of the Union." Opposed to Clay, who in 1844 was nominated for the Presidency by the Whigs, Polk was nominated on a strictly annexation platform; and he was elected. Calhoun was made his Secretary of State. Annexation measures had been already rushed through Congress, and as a sop to the North it was decreed that all the teritory of Texas north of 36° 30′ should be allotted to free labor. But Texas owned not one acre north of that line. It became therefore a later part of the Annexation scheme to steal a large slice of Mexico; even at the cost of war. General Taylor was ordered with all our disposable army to march to the Rio Grande. The war that was prepared for soon followed; and the United States had the glory of whipping a weaker neighbor, while robbing her. It made the blackest chapter in our history.

Now came the beginning of the final struggle for sec-

tionalism. Mexico had abolished slavery. Annexing
her territory added therefore to the free labor section.
Calhoun met the point with this constitutional interpretation ; that " every citizen of any State had a right
to migrate into any territory of the Union, carrying
with him whatever the Constitution of his own State
recognized as property ; and this property must be
guarded by Federal authority wherever he might be."

The Annexation scheme increased the number of
Northern abolitionists. The war revolted a large
number more of the better element of conservatives.
This change of sentiment was telling in Congress.
The time had come, thanks to Mr. Calhoun, for a new
phase of the whole battle of sections. The Wilmot
Proviso of 1846 provided as a condition to the acquisition of any territory from Mexico, that " slavery shall
never exist in any part, except for crime." In 1847 a
bill providing a permanent government for Oregon,
New Mexico, and California, embodied a reference of
the admission of slavery to the Supreme Court. The
Court was known to be sure to decide for slavery.
Almost immediately the Senate reported and passed a
bill to abolish slavery in the District of Columbia.
The House refused to concur. But at this point the
result of a geographical struggle was plainly fore-
shadowed. All the Southern Senators left their seats
in the Capitol and organized elsewhere. They drafted
a passionate appeal to their constituents ; but went
peaceably back to their seats after the action of the
House. Oregon was admitted as a free-labor State.
But during the debate over New Mexico, and California, Mr. Douglas moved " that the line of 36° 30',
known as the Missouri Compromise line, approved
March 6, 1820, be and the same is hereby declared to

extend to the Pacific Ocean "; and that all future
territorial organization should follow the rule then
adopted. This passed in the Senate but failed in the
House. The experiment as first applied had proved
anything but pacific and unifying ; what could have
come of its extension but increased sectionalism ? The
contest must be fought to an end ; one section must
overcome the other so far as social contrasts were concerned ; the sooner the better.

The last decade of the furious conflict began in 1850.
No one could foretell that its end would be in ten
years ; but all could see it was now " irrepressible."
William H. Seward's definition of the contest at this
stage was " Slavery regards disunion as among the
means of defense ; Freedom maintains the Union of the
States one and inseparable ; now and forever." He
did not add that secession had been a favorite Northern
means of defense ; and that at that very time extremists
were advocating separation from the South, as the
shortest way out of the difficulty. Northern sentiment
was crystallizing about the declaration " not one foot
more of slave-labor territory." " No more territory
unless it be free," voted the New York Whig Convention in 1847. In 1849 the Democratic Convention of
the same State voted, " We are opposed to the extension of slavery." The heat of the controversy consumed
all other questions. Secession was a daily threat.
Mr. Clay turned once more to his favorite rôle of compromises. It was just thirty years since the first great
compromise of 1820 had been adopted. It had from
first to last been a promoter of schism and mischief.
The gist of the new proposition was, first, admit California as a new State ; second, leave New Mexico to
organize without restriction ; third, abstain from abol-

ishing slavery in the District of Columbia ; fourth, prohibit the slave trade in the same district ; fifth, enact
more stringent measures to restore fugitive slaves ;
sixth, deny the right to meddle with interstate slave
trade.

This was dubbed the Omnibus Bill ; and it met at
once the opposition of President Taylor and his cabinet.
The struggle with Taylor was desperation itself.
Southern Whigs went to him demanding his aid. He
replied he wished California admitted on her own
merits, and no trick to admit more slave territory.
But would he refuse to sign any bill debarring slavery ?
He would, he replied, sign any constitutional bill.
They then threatened secession. He answered, " I
will command the army in person, and any man who
is taken in treasonable acts I will hang as I hung
usurpers at Monterey." He then sent for Hamlin of
Maine, and requested him to stand firm, " for," he
added, " this Omnibus Bill means Disunion, and Disunion is treason.   .   .   .   I will see that traitors get
their deserts."

Clay was asked when he would consent to a dissolution of the Union. " Never," he replied, " never for
any possible contingency, unless Congress usurps the
power to abolish slavery in the States where it exists ;
and that I am sure it will never do." " Even," he
cried, " if my own State, contrary to her duty, should
raise the standard of Disunion against the residue of
the Union, I would go against old Kentucky, much as
I love her. If," he added, " the people of any State
choose to place themselves in military array against the
government of the Union I am for trying the strength
of the government."

But the President suddenly sickened and died. Clay

renewed the struggle as the great hope of his life. It
was to be his final act. Webster joined his great col-
league- in an effort to save the Union, or, as it was
asserted, to win the Presidency. He was not for concession to slavery, he declared ; but he would sustain
the constitutional rights of all sections. " I mean," he
said, " to stand on the Constitution ; I need no other
platform. I shall know but one country. I was born
an American, I shall die an American." History will
not fail to write that Webster was never more sincere,
never more eloquent. Emerson gave him a just criticism when he said that Webster never had believed in
self-government, but in government. He was at the
close of his political life what he was at the beginning
of it, a Federalist. Like Clay he insisted that the only
question of the day was, Are you for the Union ; not
Are you a Whig or Democrat ; and all attempts at secession must be promptly put down. Abolition agitation
at the North, and Disunion agitation at the South, were
condemned with equal force. He would put a stop to
everything that irritated and alienated. It was too
late. Mr. Webster was himself ground between the
millstones of sectionalism ; and hurried, a misunderstood and disappointed man, to his grave.

Calhoun's speech was not novel;· it was not great; it
was an echo of 1832. It laid down an ultimatum. It
hoped for the same victory as in 1832. He said the
North must give equal rights to the South in the Territories. She must send back fugitive slaves ; and
must stop all agitation; she must consent to a Constitutional amendment which would restore the balance
of power between the sections. This constitutional
amendment he elsewhere explained must be two co-
Presidents ; elected one from the North the other from

the South, and each to have a veto on all national legislation.

Clay was so feeble that he leaned on a friend to enter the Hall. His speech was pathetic and deeply earnest. He had an anxious desire he said, to present the olive branch to both parts of this distracted and at the present moment unhappy country. " Two months ago," he added, " all was calm; all now is uproar, confusion, and menace to the existence of the Union." Benton laughed at the threats of Disunion, as too trivial for serious notice. Chase spoke of the " stale cry of Disunion."

It remained for the intensifying abolition sentiment of the North to be heard. William H. Seward had said in Cleveland in 1848, " Slavery can be limited to its present bounds ; it can be ameliorated ; it can and must be abolished ; and you and I can and must do it." While Webster's first thought was preservation of the Constitution and Union, slavery or no slavery, Seward's first intent was to get rid of slavery in order to preserve the Union. He said, " The question of dissolving the Union is a complex question ; it embraces the fearful issue whether the Union shall stand, and slavery, under the steady peaceful action of moral, social, and political causes be removed by gradual voluntary efforts ; or whether the Union shall be dissolved in civil war, bringing on violent but complete and immediate emancipation." In regard to the fugitive-slave question he added, " I say to the Slave States you are entitled to no more stringent laws ; and such laws would be useless. Has any government ever succeeded in changing the moral conviction of its subjects by force ? " The words of Seward as belonging to the forward-lookers and upward-lookers were more important than those

of Webster and Calhoun and Clay. When he came
to the territorial question he said, " The Constitution
regulates our stewardship ; the Constitution devotes
the domain to union, to justice, and to liberty. But
there is a law higher than the Constitution which regulates our authority over the domain, and devotes it to
the same noble purposes."

It was a battle of giants such as this country never
had before seen in Congress ; and to which nothing
seems likely again to be comparable. It reached back
for eloquence, pathos, and greatness to the Continental
Congress in which was enacted the drama of the Revolution. That the South was steadily growing into a
conviction of being deprived of its Constitutional rights
is unquestionable. The North was also steadily growing into a conviction that the Southern interpretation
of such rights compelled it to forego its moral convictions.

The Compromise of Mr. Clay was finally adopted.
But it had no further effect than to shift the form of
the struggle. A fugitive-slave law followed as a necessity. Its main principle was no doubt constitutional ; its details were not ; yet it was impossible to
execute it. The time was near when the Constitution
must be amended, or be overthrown. It was not any
longer the recognized law of the land. The " higher
law," first promulgated as we have seen in the South,
was now the Golden Rule of the North. The fugitive-
slave law provided for the surrender to the claimant of
each alleged fugitive from the South, without allowing
trial by jury. " In no trial or hearing under this act
shall the testimony of such alleged fugitive be admitted
in evidence." Marshals charged with the duty of
arresting fugitives, were empowered to summon to

their aid any bystander. Three-fourths of the people
of the North stood ready to refuse to obey. Several
States passed nullification acts. In Wisconsin the
Supreme Court decided the fugitive-slave law unconstitutional. The United States Supreme Court overruled
this decision ; but was not obeyed, and its overruling
could not be enforced. Tragedies multiplied. Fugitives committed suicide ; and in one horrible case a
large company killed all their babes and small children.

That the question was finally one of migration and
colonization was apparent. Slavery north of a certain
line was forbidden ; freedom south of that line was not
forbidden. Such a division of the soil was preposterous.
Up to 1830 the two peoples had marched with almost
equal persistence through the wilderness. But the
Yankee always went to civilize ; the poor Southerner
loved the wilderness. The former went with the
plough ; and the first buildings were a school house
and a mill. As soon as possible he organized a church,
and then a town meeting. The lack of equality in
Southern society made such a gathering a disagreeable
if not quite useless affair. The North pushed turnpikes and canals, and soon after railroads always to the
front.

Back of the township lay the family in historic evolution. The New England family was showing as yet no
signs of disintegration. It was a wonderfully self-com-
pleting unit. Every boy was taught not only to till the
soil but to work with craftsmen's tools. That was a
rare man who was unable to make his own shoes, build
his own houses, butcher his own meat, build a good
stone wall, and do a dozen other artisan specialties—
and do them well. Every woman was expected not
only to make her own butter and cheese, but to spin her

own home-grown wool, weave it into cloth, and make it into all needed household garments. She also supplied the house with soap and candles, carpets and bedding; besides being abundantly able when needed to milk the cows, or drive oxen, or split and saw wood. With the town meeting always in sight, the Northern migrant naturally thought less of the State, and more of the general Union. It was hard for him to comprehend the promptness with which the Southerner said, "I must go with my State, right or wrong." He could comprehend the toast, "The Union right or wrong."

By 1850 the Northern migrants had touched the Rocky Mountains; had gained Oregon, California, and a good share of what had been stolen from Mexico. The South had crossed the Mississippi barely and added Texas. The struggle of sections, as two equal parts of the United States, was settled already; but Southern leaders refused to see the fact. They proposed to force slavery over into new territories. Webster did not see that Congress need legislate on the subject, for " God's decree had fixed the future of the Southwest." Slavery could not establish itself there in the nature of things. The two forms of labor had had a fair competitive struggle. Free labor could outwork, outsettle, outvote slave labor. The country was no longer South and North; but Northeast and Southeast, Northwest and Southwest. Slavery held but one of the four sections. This was felt by the Old South more than by the Old North. The former was tied up to cotton and slavery. It could not build factories; for free labor would not go there to run them; and slave labor could not. Could conflict now have been held off for ten years, secession and war would never have occurred.

But this delay could not have been obtained, even had the combatants comprehended the future. Organic efforts for secession were made in 1851. It was hoped by Southern fire-eaters in South Carolina, Georgia, and Alabama, and Mississippi to bring all the Southern States into unison. The immediate result was acquiescence in compromise; except in Mississippi. There Quitman led a distinctively disunion party, as candidate for governor. He was thoroughly defeated by Foote, who had been nominated by the Unionists. Jefferson Davis at this juncture assumed the leadership of Southern extremists; and on his shoulders was supposed to rest the mantle of Calhoun. With unbending rigidity, unflinching courage, and high moral character, he had consecrated himself to the South. Underneath his affection for slavery was his devout adhesion to State rights. Combining these two attachments, he affirmed as cardinal that slavery was not a local institution, but a United States institution; and had a right anywhere under our flag. But if denied such equal privilege, any State might withdraw from the Union, as its sponsor. Calhoun had said, " If you ask me the word which I wish engraven on my tombstone it is Nullification." Mr. Davis preferred the word Secession. Sumner led the Northern advance guard with, " Freedom and not slavery is national"; Davis led the aggressives of the South with, " Slavery is national and not sectional." The battle was at last between these two principles.

It was by force of this final alignment of sectionalism to give up sectionalism altogether, that in 1854 Stephen Arnold Douglas introduced into the Senate a Bill to abolish the old Missouri Compromise of 1820. Congress had refused to agree with his proposition to extend the

line of 36° 30' to the Pacific; he would now blot
that line out as a divider of the people into two
sections. This would admit that slave labor and free
labor were henceforth on equal terms everywhere in
national territory. The North remembered that Mr.
Douglas bore the name of the traitor Arnold. The
South was of much the same mind. The Bill was carried through Congress. The era of labor compromise
had lasted from 1820 to 1854, but it was now ended.
The North was furious ; the South was only temporarily
jubilant. It knew very well that nothing substantial
was gained. On the contrary there was a vast increase
of anti-slavery conviction at the North. The feeling
grew that slavery did not wish for peace ; that it had
no limit to its demands. In 1849 Douglas himself had
said that " the Missouri Compromise had become
canonized in the hearts of the American people, as a
sacred thing, which no ruthless hand would ever be
reckless enough to disturb." His own hand had now
done the sacrilegious act. Sumner saw the consequences. " It annuls all past compromises with
slavery," he said, " and makes all future compromises
impossible."

Meanwhile efforts to enforce the Fugitive-Slave Law
aggravated and concentrated Northern sentiment. A
comic act of diplomacy was enacted at Ostend in Belgium, by our ministers to England, France, and Spain
—Buchanan, Mason, and Soulé, who published a proposition that we should purchase Cuba ; and Spain refusing to sell, we should steal the island. The intent
was to increase slave territory.

Lyman Beecher, who by a remarkable series of dis-
courses had, about 1820, roused a universal temperance
sentiment at the North, that in ten years swept out of

existence ten thousand stills, and every sideboard in
Northern Christian homes, was on the subject of slavery
conservative.   But he showed the virility of his genius
by begetting a race of giants that became the mightiest
sectional forces of the whole era.   Henry Ward Beecher
turned his pulpit into a rostrum for abolition ; and his
sister Harriet wrote the book that shook slave labor to
the foundations.

With no guarantee against slavery north of 36° 30′,
freedom claimed equal privilege south of that line.
Massachusetts deliberately bent the line of migration
southward.   The New England Emigration Aid Society was formed to pour free-soil settlers into Kansas,
fast enough to control the vote that should organize its
government.   The South was nearer Kansas ; the
North overwhelmingly more populous and wealthy.   It
had no members of society that it dared not take upon
contested ground.   The South was quick to feel its disadvantage, and appeal to force.   The North was cool
and resolute, and met force with force.   Beecher's
church contributed rifles as well as bibles.   The sentiment against meddling with slavery where it existed
did not object to forbidding its extension.   Free-State
men denied that they were interfering with Southern
institutions ; they claimed their right to settle home-
steads.   The subject of labor must be settled by the
majority.   It was an amazing movement.   " Free
soil," and " Free State " were taking terms, where
" Abolition " was detested.   The old Liberty party
had never rallied but 156,000 voters ; the new Republican party was launched 1,341,264 strong.

The Kansas war was not the wrestle of two sections
for predominance ; it was the struggle of slavery for
permission to exist at all.   This was felt by all parties;

but not yet was it formalized into a political creed. The
Buchanan administration certainly sympathized with
the South. The contest ended in making a free State,
although it was not at once admitted as such into the
Union. Slavery was hemmed in. No more slave territory could be created. There was but this choice left
for the South, either to yield the basic institution of its
social life, or leave the Union. Calhoun had said,
" The relation which now exists between two races in
the slaveholding States has existed for two centuries.
It has grown with our growth and strengthened with
our strength. It has entered into and modified all our
institutions, civil and political. We will not, we cannot permit it to be destroyed. Come what will, should
it cost every drop of blood and every cent of property,
we must defend ourselves. We should stand justified
by all laws human and divine." Alexander Stephens
said, " This stone which the builders rejected is become
the chief stone at the corner of our new edifice." Jefferson in 1821 wrote, " We have a wolf by the ears, and
can neither hold him nor let him go." Madison in
1835 mournfully said with regard to slavery, " I own
myself to be in despair almost." He believed the
blacks must ultimately go somewhere, but where was
the unsolved puzzle. The leaders of secession had a
real and not a fancied dilemma. It was useless for
Northern leaders to say, We do not propose to meddle
with slavery in the States where it exists. To prevent
slavery from expansion collateral with free society was
to doom it to self-expungement.

The doctrine of a law above the Constitution was
bringing forth fruit at the North. South Carolina first
injected it into our political life, but John Adams as
early as 1768 in the first Continental Congress said:

" You have rights antecedent to all earthly government ; rights that cannot be repealed or restrained by
human laws ; rights derived from the great legislator
of the universe." Dr. Channing also stated the conviction of New England conscience when he said of the
Fugitive-Slave Bill, " A higher law than the Constitution protests against this act of Congress. According
to the law of nature no greater crime against a human
being can be committed than to make him a slave."
Wendell Phillips plucked fruit from this tree when he
faced a Boston mob. " Governments exist," he said,
" to protect the rights of minorities. We have praised
our Union for seventy years. This is the first time it is
tested. Has it educated men who know their rights,
and dare to maintain them ? Can it bear the discussion
of a great national sin, anchored deep in the prejudices
and interests of millions ? If so it deserves to live. If
not, the sooner it vanishes out of the way the better."
When the South finally began to call conventions for
taking measures to get out of the Union, Mr. Phillips
argued that they should be bid to go. He said: " In
my soul I believe that dissolution of the Union, sure to
result speedily in the abolition of slavery, would be a
lesser evil than the slow faltering disease ; the gradual
dying out of slavery, constantly poisoning us. All
hail ! then disunion. The sods of Bunker Hill shall
be greener now that their great purpose is accomplished. Sleep in peace, Martyr of Harper's Ferry !
your life was not given in vain. Soon throughout
America there shall be neither power nor wish to hold
a slave." The numbers were growing at the North
who were ready to say, As between perpetual slavery
with the consequences, and separation, let us separate.
Some leaders had been found to argue that the North

would thrive better industrially as well as morally by withdrawing from the Union. Retaliating, the South offered to ally itself with the Middle and Western States, and leave out New England.

Up to this time the great political parties had kept up a form of nominating their tickets across the sectional line. That line was now gone. In 1856 the new Republican party, nominating Fremont and Dayton, was met with the outcry that its ticket was sectional. Buchanan was besieged with letters stating the case in that light. Lacking the strong manhood to denounce threats of secession, he was, long before his election, snarled up in the meshes of complicity. He could not afterward take a Jackson stand, and sustain the Union. He made it an electioneering card that he ought to have votes to save the Union. Governor Wise wrote that the Southern States would not submit to a sectional electional. Buchanan wrote that his advises were that the election of Fremont would involve " the dissolution of the Union, and that immediately." It was sectionalism. The whole of the United States from Maine clear around to California enclosed the old South, and linked hands with Kansas on the way. Longfellow, Bryant, Emerson, made political speeches. Fillmore, like Buchanan, wished to make political capital of the situation ; saying: " We see a political party for the first time selecting candidates from the free States alone. Can they have the madness to believe that our Southern brethren will submit to be governed by such magistrates ? "

But Buchanan and Fillmore were both Northern. They would not be able to more than defer secession, on their own showing. The fact was, the Old South must secede, or remain in the Union with the certainty that slavery was to be eliminated. The new Republican party had been organized to protest against the invasion of State rights. The chief emphasis of the platform had been laid on the fact that the people of Kansas had been fraudulently and violently robbed of the rights guaranteed them by English common law ; and confirmed by the Constitution. Their territory had been invaded by armed forces ; spurious legislative, judicial, and executive officers set over them ; test oaths of an extraordinary nature had been imposed as a condition of exercising the right of suffrage and holding office; the right of jury had been denied ; the right to be secure in their persons, houses, papers, and effects, against unreasonable searches, had been violated ; they had been deprived of life, liberty, and property without due process of law ; the freedom of the press and of speech had been abridged ; the right to choose representatives had been made of no effect ; this, slightly abridged, constituted the arraignment of the party in power—and for this high crime against " State rights " the new Republican party entered the field, " to bring the actual perpetrators to a sure and condign punishment."

But four years later the same party, while not waiving its State-rights doctrine, became peculiarly and distinctively devoted to the sustenance of the national Union of the States. " The rights of the States and the Union of the States must and shall be preserved." The chief article of the platform read that " to the Union of the States this nation owes its unprecedented increase in population, its surprising development of material resources, its rapid augmentation of wealth, its happiness at home and its honor abroad ; and we hold in abhorrence all schemes for disunion, come from

23

whatever source they may ; and we denounce those threats of disunion, in case of a popular overthrow of the ascendency of the Democratic party, as denying the vital principles of free government, and· as an avowal of contemplated treason, which it is the imperative duty of an indignant people sternly to rebuke and forever silence.''

The election of Lincoln in 1860 was followed by the ordinance of secession in South Carolina, December 20th of the same year.   The document was a long one ; but the sum of it was in the affirmation that '' the ends for which this government was instituted have been defeated ; and the government itself has been made destructive of them by the action of the non-slaveholding States.   Those States have assumed the right of deciding upon the propriety of our domestic institutions, and have denied the rights of property established in fifteen of the States, and recognized by the Constitution ; they have denounced as sinful the institution of slavery.''   It was certainly anticipating events to aver that the government itself was to be destructive of slavery, in any manner that had not been previously true.

Lincoln was in no sense an anti-slavery candidate. He appeared before the people as an unqualified federal Unionist.   '' I would,'' he said, '' save the Union.   I would save it in the shortest way under the Constitution. The sooner the national authority can be restored the sooner the Union will be the ' Union as it was.'   If there be those who would not save the Union unless they could at the same time save slavery, I do not agree with them.   My paramount object is to save the Union, and not either save or destroy slavery.''

Lincoln wrote to Alexander H. Stephens, who was

held to be the brainiest man of the South, asking him
if Southerners really feared that slavery would be med-
dled with in the States where it existed. Stephens replied that the real trouble was that the South resented
the moral opprobrium which the North heaped on their
social life.

Slowly and inevitably the South had been made over
to its own stronger social institution. Its sentiments
and habits of speech and action were involuntarily adjusted to slavery. Steadily also its whole intellectual
and moral life was permeated with its social life.
There was but one theme to be considered. The moral
force of the States was exhausted in questions of suppressing insurrection ; of improving slave breeding ;
of extending slave territory; of defending the ethics of
slaveholding ; of recouping runaways. The people
bristled with irritability on one subject.

The training of the North was exactly the opposite.
Only for the constant presence of the moral issue raised
by slavery it would have become a devotee of gold.
Wealth was accumulating enormously ; and the era of
its centralization in fewer hands, in larger amounts,
had come in with steam. But the North was a hundred
years farther away from the duel and the bowie knife
and the bludgeon than the South. It had come to a
strong reliance on argument. History when justly
written will take this more into account while judging
President Buchanan. Eminently a statesman, trained
to diplomacy, he was an extreme result of the Northern
school. Afraid of bluster, he could not have comprehended such a man as Jackson. Four-fifths of the public men of the North were of the same temper. When
the clash came, the old and despised, but well-trained
abolitionists came to the front as the only qualified

leaders ; while at the South every one was retired behind the most positive secessionists.

The diverse characteristics of the South and North were manifested down to the firing on Fort Sumter. The North made no preparations for war. Its instincts were so strongly for peace that it did not yet believe war possible. Each party misunderstood the other. Hot and testy, the Southerner sneered at the Northerner as a mere mechanic or a trader, without more pluck than a yardstick. The Northerner in turn had become too much used to truculency. Randolph's mental photography anticipated a race of dough-faced neutrals in politics. But when the fight was on and the appeal was finally made to the rifle, there is no one who will say the sections were not equally matched in courage.

Lincoln was a cross of the North and the South. Born in Kentucky; bred in Illinois; he was a providential product, because he understood both parties, and had not a little of the sentiment of both sections. The solid Union was typified in his genius ; his instincts were not partisan. On his way to Washington he said, "There will be no bloodshed unless it is forced on the Government ; and then it will act only in self-defence." There is nothing in his career to show that Mr. Lincoln could not have been trusted by the South as completely as by the North. Not only was Lincoln totally averse to meddling with slavery, the Supreme Court was positively and aggressively pro-slavery ; while Congress, no longer as yielding as formerly, had invariably passed compromise measures ; and up to the hour of secession it had refused to admit Kansas with a Free-State Constitution.

The fact was Mr. Davis was too intently a one idea

man ; incapable of broad generalizations ; and as incapable of the slightest flexibility.   Secession was illogical as a means of betterment ; and Mr. Stephens plainly told them so.   But from the standpoint of slavery as a national principle, and therefore State rights beyond State lines, secession was logical.   Sectionalism and union had been tried ; fully tried, and had proved a failure.   Both sides must now struggle for State rights and nationalism.   The North declared for State rights and the whole Union ; the South for State rights and a divided Union.

It was unfortunate for the South, but not unfortunate for humanity, that the principle of State rights and reserved integrity became so integrated with the preservation of slavery.   The principle which in Jefferson's day contested the enactment of tyrannous sedition laws, and in 1832 aimed to nullify the enforcement of sectional taxation, and the fostering of local industries at the general expense, had in 1861 sunk into a struggle to perpetuate and extend an institution that Jefferson abhorred, and South Carolina when she entered the Union condemned as barbarous.   But this was really what had come about.   It is pitiful to read the South Carolina Ordinance to dissolve her union with the United States. After a thoroughly statesmanlike exposition of the sovereignty and freedom of each State, demonstrated by history, the whole argument slumps into the single and only grievance that an " Increasing hostility on the part of the non-slaveholding States to the institution of slavery has led to a disregard of their obligations."   These obligations being the " rendition by the several States of fugitive slaves."   Thus the Constitutional compact had been deliberately broken and disregarded by the non-slaveholding States ; and the consequence followed that South Carolina was released from her obligation. In other words, the several non-slave-holding States, by exercise of that very State-sovereignty which was so sacred to the South, had refused to engage in catching runaway slaves to be returned to bondage. Acts to this effect were passed by fourteen States, in obedience to the moral sentiment of the people and of the whole civilized world. They had not been passed until acts of brutality had shamed the most conservative, and enraged the liberal. Marshals had broken open the domiciles of freemen to hunt for negro fugitives; and slave hunters had shot runaways in the streets. Personal-liberty bills may have been extra-Constitutional; but that did not relieve the shame of an act of secession based on no other alleged offense. It was the glorious principle of State rights for which the South had stood conspicuous, brought into temporary subjection to a vast social disease. Slavery had become precisely what Madison feared, a moral influence to pervert the manliest sentiments, and render ignoble the truest chivalry of America.

The struggle to prevent dissolution was pathetic. Mr. Powell of Kentucky proposed a committee of thirteen, equal in number to the original States, to consider the state of the Union. Johnson of Tennessee proposed an amendment to the Constitution, electing President and Vice-President by direct vote of the people; while alternately the President should be from either side of the old sectional line. Calhoun's idea of two Presidents, one from each section, was revamped. Johnson declared there was no cause for disunion. "Mr. Lincoln was a minority President; but he should not run for that. He was in the Union and meant to stay in it." Mr. Crittenden of Kentucky offered a compromise to restore the old line of 36° 30'—prohibiting slavery to the north of it, and leaving its existence to the south to the act of each individual State. Thurlow Weed, in his Albany *Evening Journal*, proposed, (1) Restoration of the Missouri Compromise ; (2) Instead of returning fugitive slaves, give the North the option of paying for them. Charles Francis Adams made an unassailable point when he asked, Who excluded slaveholders and their slaves from the territories? The Supreme Court had decided for them. No majority in Congress was against any such freedom. He said, " They will not leave the rich bottom lands with the cotton plant to go to a comparatively arid region farther off." The fact was slavery could not do the territorial work of free labor—the pioneering and settling and subduing. But it was not in human nature to confess it.

Douglas bravely went through the Southern States, arguing against secession. It is a mistake that Davis urged hasty action on the the part of the South. He plead for consideration and delay. He insisted that he spent his nights in prayer. Wade of Ohio said he " did not so much blame the South ; for they had been led to believe the dominant party was ready to trample their institutions under foot." Virginia proposed a conference of the States to be held at Washington, to consider and agree upon some suitable adjustment of difficulties. The Conference was attended by six slaveholding States and all but five of the free-labor States. Both sides, however, were evidently a little more ready to fight than to yield. The Amendments to the Constitution suggested were much like the Crittenden compromise resolution ; and were carried by a majority vote.

The adoption of the Crittenden Compromise would have averted immediate action on the part of all States

but South Carolina ; but it would have driven the
North into revolt. For the time being the conceding
element was at the front, but there was not the least
weakening of popular sentiment as to slavery. August
Belmont tells of a meeting in New York of such men as
Astor, Aspinwall, Grinnell, Hamilton Fish, etc. ; and
that they were unanimous that the North should take
first steps for reconciliation—but these men represented
the trade, not the conscience of the North. Seward
wavered—believing that while the Republican party
was now uncompromising it might be more pliable after
a while. Lincoln also hesitated—but he was a logician.
His trouble was that he foresaw the danger of keeping
the government together by force. " Ours," he said,
" should be a government of fraternity." All he foresaw of mischief to come from war power has come to
pass. But he finally concluded, precisely as Jefferson
did, that to recreate a geographical line to divide sectional ideas would only perpetuate mischief. " Let that
be done, and immediately filibustering recommences."
This was a general but not formalized conviction of the
majority on both sides ; sooner or later we must fight
this out—let the struggle come at once. The common
people of the North were tired of hearing the threat
Disunion, Secession. They felt, Let it be tried. On
the other hand the South knew well that their domestic
institution was despised by the North, and that no sort
of Constitutional safeguards would prevent this from
being shown.

Davis echoed Stephens's opinion when he said, The
true remedy for our ills is " in the hearts of the people."
Lowell retorted, " Our very thoughts are a menace."
Davis was not wholly wrong when he added, " Give us
evidence that mutual kindness shall be the animating

motive, and then we may look hopefully for remedies—
not by organizing armies." The fact was the North
had grown into a state of hatred for the South—and it
was fully reciprocated. War existed in the people's
hearts.

Crittenden asked January 3d, "Are we prepared for
war—are we prepared in our hearts for war with our
own kindred and brothers?" The North felt that it
was. The South was not far from the same conviction.
Yet the Crittenden Compromise would probably have
been accepted by a majority of the people if it had come
to a popular vote ; with however a deep reserved belief
that it would not end the struggle, or get rid of the
source of danger ; that it was a mere makeshift. Senator Chase inflexibly opposed all compromise, as sure
only to lead to deeper difficulty. Chandler wanted
" stiff-backed men " sent to the Washington Conference; adding heartlessly that " without a little blood-
letting this Union will not in my opinion be worth a
rush."

At the meeting held November 12th, 1860, at
Charleston, it was resolved, " Our Confederacy must
be a slaveholding confederacy. We have had enough
of a Confederacy with dissimilar institutions." Lincoln
had already expressed the same convictions from a
Northern standpoint, and been bitterly assailed for it
by the South.

The Crittenden Compromise measures did not pass,
and the committee of thirteen was a failure. Seward
wrote, " The Republican party is to-day as uncompromising as the Secessionists in South Carolina." Of the
debates in the senate he wrote " they are presumptuous
on the part of the South, feeble and frivolous on the
part of the North." The drift was too strong for states-

manship. Compromise had had its day. But Congress
did, on Feb. 27, 1861, by a two-thirds vote, propose as
a XIII Amendment to the Constitution, that no amendment should be permissible to the Constitution, allowing Congress to meddle with domestic institutions of
any State; or to abolish slavery where it then existed.
This vote had no effect whatever on the course of
events.

Scott, as commander-in-chief, wrote to President
Buchanan that the nine seacoast forts should be immediately garrisoned. The garrulous general added that
the South had better be allowed to peaceably secede.
Buchanan wobbled between the two positions, and did
nothing. Under similar conditions Jackson had said,
"Execute your views at once—you have my carte
blanche in respect to troops—the vessels shall be there."
Buchanan pleaded danger of farther irritating the
South ; and left revolt to have full swing. He sat in
the White House with Jefferson Davis, discussing his
message ; and mildly obeying suggestions. This at
least can be said for the gentlemanly statesman, that he
denied the right of secession (while denying the right
of the government to stop it). He was hopelessly
snarled in the meshes of logic, and action could not be
thought of till he had unknotted his Constitutional
questions.

The appeal to force was simply inevitable. All conditions compelled it. There was at last a general feeling that Providence ordered it. On both sides praying
to the God of battles was equally vigorous and equally
sincere. North and South had alike become heart-sick
of cowardice. Courage of opinions and consistency
were honored. When the war broke out Davis was
positively popular at the North, and Douglas at the

South. The North hated Toombs, but respected him.
It was the men who were making political capital out
of sectionalism who were detested. John Brown was
resurrected by popular sentiment, and set to marching
at the head of the Northern forces.

That the bulk of the Southern people desired to leave
the Union is not true. In North Carolina the first
State convention was overwhelmingly loyal. The
State was carried out by trickery. Virginia also at
first refused to secede, while Eastern Tennessee, and
what became West Virginia, were as loyal as Michigan
and Massachusetts. Notwithstanding the profession of
State rights as the basis of their withdrawal from the
Union, the Confederate States recognized no such rights
within their limits. East Tennessee was harried and
dragooned. Thousands of its people were arrested and
sent to prisons in Alabama ; while other thousands fled
into the free States. Due credit has never been given
to the Southern border States. By a four-fifths vote
the Legislature of Kentucky ordered the national flag
hoisted over the State House at Frankfort as a reply to
the invasion of the Confederates ; and in September of
1861 the State declared war on the Confederacy. No
State occupied a position of greater peril ; and none
conducted itself more judiciously and loyally. Of those
who voted for secession, it is probable that we can say,
as Jefferson said of the Federalists in 1814, they did
not intend to stay out. The dominant purpose was to
secure a vantage ground for dictating terms. Judge
Hand, addressing a large audience in Maryland, urged
a union " of the entire South to form a compact until
they could be guaranteed all Southern rights, and that
their institutions would be respected. They could afterwards, in solid phalanx, or separately, present an ultim-

atum to the North, and return if practicable, with the
present Constitution properly amended, on amicable
terms." Stephens said that Georgia was won to secession by the cry " Better terms can be made out of the
Union than in it." Stephens own course can be understood only from this standpoint.

No thorough student of the period can fail to be
grateful that in this shoreless floating of the statesmen
of the period, Lincoln had a clearly defined policy.
The vacillation of his predecessor aroused unwarranted
hopes, and aggravated disappointment. Whatever
momentous responsibility he undertook to assume, he
first told to his cabinet, then to a tattling circle of un-
official advisers, then to all the world, winding up by
concluding not to do anything. Striving to please
everybody he lost every friend. Even General Cass
left the cabinet in mingled despair and disgust. The
old gentleman boasted in a letter that " all the trouble
had not cost him an hour's sleep. He left it to Providence." But the sleepless people believed him to be
in his dotage. When crowded by the South Carolina
Commissioners to send Anderson back to Fort Moultrie,
he replied, " You press me too importunately. I always
say my prayers when required to act upon any great
State affair." Public opinion of Lincoln during 1861
was no less unfavorable. Bowles of the Springfield
Republican wrote to Dawes, " Lincoln is a Simple
Susan." After the battle of Bull Run, Chase said to
Asa Mahan, " If Lincoln lives the North is whipped."
It was the general belief that he was incompetent to
lead in such a crisis. His inaugural was temperate,
and as far as possible from provocative threats. " In
maintaining the Union I shall use the power confided
to me to hold, occupy and possess the property and

places belonging to the government, and collect the duties and imposts ; but beyond what may be necessary for these objects there will be no invasion, no using of force against or among people anywhere.'' The people did not as yet understand him.

The affair was out of the hands of Congress, out of the hands of the Executive—it had long been out of the hands of the Supreme Court ; the appeal had been made to the people.   Sumner wrote, '' The question will soon be carried before that high tribunal, supreme over Senate and Court, where the judges will be counted by millions; and where the judgment rendered will be the solemn charge of an aroused people.'' Again and again in our history we have seen that, while the people create bodies of representatives for legislative, or judicial, or executive convenience, in all great crises the people do assume direct control of affairs.   The ballot failing, resort was taken to the bullet.   The Constitution was for a while allowed to hinder martial affairs ; after which it was coolly ignored. The functions of the Supreme Court were suspended. Congress as well as the President refused to consider the opinions of the Judiciary.   Congress and the President also conflicted, with varying success.   This conflict culminated under Lincoln's successor in an effort to impeach and remove ; after which the Constitutional machinery began to resume its methodical action.

The South and the North entered the conflict as we have seen with equal intent to sustain State rights. They equally at once yielded every vestige of this principle.   The South acted in all ways as a centralized unit.   The North came out of the war more centralized than even during the period of Federalism under John Adams.   This drift has continued to the present day.

There had been a notable suggestion all through the controversy from 1820 to 1860 that the original Union was a compact entered into between North and South to sustain a balance of power. This was a perversion of history. The States acted each by itself. The first to accept the Constitution was a Northern State. The last to approve was also a Northern State. The sections were created by the Missouri Compromise. Slavery destroyed State rights and created section rights. The final question was not whether a State could secede, but whether a section could withdraw. Not a State would have seceded if convinced it must stand alone. The height of absurdity would have been South Carolina paddling its own canoe as a nation. The Confederacy at once conducted itself as a compact body.

To summarize this final attempt at secession in the United States, we find it to have been the result of the drawing of a geographical line, on opposite sides of which the attempt was made to sustain diverse and conflicting forms of labor. There grew up institutions and customs, theories, and sentiments so entirely opposite that opposition must express itself with constantly increasing bitterness. Separation had been suggested on either side with about equal frequency ; Nullification of Congressional acts had been' undertaken on both sides ; nor had the Supreme Court been held by either North or South as final arbiter. State rights had been affirmed by Northern as well as by Southern States, and there was no difference as to the principles of State integrity and State sovereignty.

The North would have separated from the South had not the South undertaken to secede from the North. If, in the final struggle under the Constitution, the Fugitive-Slave Law had not been nullified in every Northern State ; had Kansas been won by the South, and the Missouri Compromise line sustained to the Pacific, a Free-State Republic would inevitably have resulted. The appeal to Congress had however resulted in more than justice to the South. An appeal to the Supreme Court was also favorable to the South; while the appeal to the Executive had been partial to the South. The final appeal to the people and to the laws of nature had been taken.

It has not been my purpose to undertake a history of the civil war ; nor even to touch the hundreds of episodes—involving no new principles—but immensely important in deciding the issue. These belong to the general historian. The end of the struggle was most admirably sketched, with the principles involved, by Grant in his proclamation to the soldiers, announcing the termination of hostilities :

" Soldiers of the army of the United States :

" By your patriotic devotion to your country in the hour of danger, your magnificent fighting, bravery, and endurance, you have maintained the supremacy of the Union and the Constitution, overthrown all armed opposition to the enforcement of the laws, and of the proclamation forever abolishing slavery, the cause and pretext of the Rebellion ; and opened the way to the rightful authorities to restore order, and inaugurate peace, on a permanent and enduring basis on every foot of American soil. Your marches, sieges, and battles, in distance, duration, resolution, and brilliancy of result, dim the lustre of the world's past military achievement, and will be the patriot's precedent in defense of liberty and right in all time to come. In obedience to

your country's call you left your homes and families,
and volunteered in her defense. Victory has crowned
your valor and secured the purpose of your patriotic
hearts ; and with the gratitude of your countrymen,
and the highest honors a great and free nation can
afford, you will soon be permitted to return to your
homes and families, conscious of having discharged the
highest duty of American citizens. To achieve these
glorious triumphs, and secure to yourselves, your
fellow-countrymen, and posterity the blessings of free
institutions, tens of thousands of your gallant comrades
have fallen and sealed the priceless legacy with their
blood. The graves of these a grateful nation bedews
with tears, honors their memories, and will ever cherish and support their stricken families.

'' (Signed)

'' U. S. GRANT, Lieutenant-General.''

On the Union side there had been in service 2,688,523
men. - Of these 56,000 died on the field ; 35,000 in
hospitals ; and 184,000 were carried off by disease.
Adding those who afterward died of wounds or disa-
bilities ; the loss was nearly 300,000. The losses on the
Disunion side were probably not far from an equal
number.

Two principles were established by the war of 1860
as fundamental and unquestionable. The first of these
was that slave labor is not compatible with the present
industrial conditions of the world, and must be obliterated. The emancipation proclamation was a war measure. It was promulgated in behalf of the Union and
not of humanity. But it was recognized as the culmi-
nating act in a drama far larger than a war of secession.
The Constitutional amendments that followed estab-

lished the principle that " neither slavery nor involuntary servitude, except as a punishment for crime, whereof the party shall have been duly convicted, shall exist
within the United States, or any place subject to their
jurisdiction " ; and the right of suffrage was permanently based on common conditions for citizens of all
colors. The Republic made a marked and emphasized
stride ahead in social and political sentiment. And we
may set down to the middle of the nineteenth century,
and especially to the Civil war, this addition to the
morals of the world, that freedom is the heritage of
every man that is born into it. There was a general
feeling that slavery was a curse, and that the war was
God's judgment on the people that had tolerated it.
Whittier sung,

> " The hand breadth cloud the sages feared,
>   Its bloody rain is dropping ;
> The poison plant the fathers spared,
>   All else is overtopping.
>   East, West, South, North,
>   It curses Earth ;
>   All justice dies,
>   And fraud and lies
>   Live only in its shadow.

> " What gives the wheat field blades of steel ?
>   What points the rebel cannon ?
> What sets the roaring rabble's heel
>   On the old star spangled pennon ?
>   What breaks the oath
>   O' men of the South ?
>   What whets the knife
>   For the Union's life ?
>   Hark the answer ! Slavery."

That slavery is a natural wrong, a curse per se, is a
rigidly modern conclusion. No Greek or Roman phi-

lanthropist so conceived it ; and the writers of both Old
and New Testaments accept it as a part of a moral
social system.   Aristotle said, " Slaves are domestic
intelligent animals."   Greek culture rested upon, and
was rendered possible only by this subjection of one
class to drudgery.   To some extent slavery in America
created a similar class of the " cultivated."   If we can
conceive of abolition as first taking place at the South,
and slavery combining with New England thrift and
logic, Boston would have had a glory to surpass
Athens.   It is hardly needful to say that no such civilization is more than temporary.   It contains the elements of self-destruction.

Negro slavery had however not been without its
ameliorating contributions to American history.   It
had secured a larger range of social sentiments, and
prevented the unchallenged dominance of Puritan con-
ceptions.   In Virginia it aided in the development of a
class devoted to statesmanship.   England and France
with their immensely older civilization had no minds to
surpass Washington, Jefferson, Marshall, Madison,
Calhoun, and Livingston.   But it was not submission
to slavery that made the greatness of these men ; it was
their antagonism to the institution.   There was a hard-
ness about New England life at the best that was lacking in the South.   Hospitality was the first of Southern
virtues ; economy the first in New England.   Channing
wrote in 1799, " I blush for my own people when I
compare the selfish prudence of a Yankee with the
generous confidence of a Virginian.   Here I find greater
vices but greater virtues than I left behind me.   Could
I only take from the Virginians their sensuality and
their slaves I should think them the greatest people in
the world."   Slavery, while depressing the poor, gave

chivalry, courage, pride, and strength of culture to
the highest class, that they could not have had otherwise.

Near the sectional line, where the conflict was incessant, developed some of the nobler, as well as some of
the meaner results of social inequality.   At Lane Theological Seminary in Cincinnati all discussion of slavery
was forbidden ; the object being to secure Southern
students.   A revolt followed ; and Oberlin became a
school for all colors as well as both sexes in 1835.   To
the existence of slavery therefore we owe the first col-
legiate institution in America where the sexes could
stand on an equality, as well as the colors.

No such vigorous altruism was ever known in the
history of mankind as that born of and nourished by
the contest with slavery.   The whole world became
agitated.   Nation after nation waked to the wrong of
holding property in man.   Equality and fraternity were
fostered.   Heroes were multiplied.   Our annals were
enriched with names we shall cherish as long as the
Republic endures.

The Southern States, no more distinctively the
South, reaped more richly than any others.   They
were rid of an institution that had been their industrial
and social curse—and it was with a sigh of relief that
the prostrate people began to look about them for re-
cuperation.   Father Ryan their chaplain poet sang,

" There is grandeur in graves—there is glory in gloom ;
For out of the gloom future brightness is born,
As after the night comes the sunrise of morn ;
And the graves of the dead with the grass overgrown,
May yet form the footstool of liberty's throne ;
And each single wreck in the war-path of might,
Shall yet be a rock in the temple of right."

A second principle equally settled by the war was that no State or States could secede from the Union at will, or by any other method than successful revolution. The course of our discussion has shown that this principle was a comparatively new one. Whatever stigma was attached to the Essex Junto, it was never suggested that if New England chose to form a separate Northern Confederacy she could not do so if she liked. While Livingston's discussion, in Jackson's name, was the first State paper defending the indissoluble nature of the Union, it was far from settling the question. In February, 1861, Greeley wrote that " If the Cotton States chose to form an independent nation they have a clear right to do so ; and if the great body of the Southern people desire to escape from the Union we will do our best to forward their views." The New York *Herald* advocated a reconstruction of the Union with New England left out. But Lincoln and Seward reaffirmed the principle of Livingston, " The Union of these States is perpetual. No State upon its own motion can lawfully get out of the Union. The laws of the Union shall be executed in all the States." Seward said, we " must change the question before the public from one of slavery or about slavery to a question of union or disunion." On that basis the war really was fought. Emancipation was brought in as "a necessity to sustain the Union." Lincoln in his inaugural said of the lawfulness of secession, " The question at issue is whether a Constitutional Republic or Democracy, a government of the people, by the same people, can or cannot maintain its territorial integrity against its own domestic foes." The abolitionists took side with Greeley. Phillips said, " Here are a series of States girding the Gulf who think their peculiar isms require a separate government. They have a right to decide that question without appealing to you or me." This was old Federalism in New England blood. The border States declared they were " for the Union " but " equally opposed to secession and coercion."

The Northern border States opposed coercion as strongly as Kentucky, Maryland, and Missouri. Hugh McCulloch says of the Indiana legislature of 1860–61, " I was astounded by the speeches of some of its most prominent members against what they called coercion of sovereign States. Indiana, they asserted, would furnish no soldiers, nor permit soldiers from other States to pass through her territory for the subjugation of the South." The people of southern Illinois were in sympathy with those of southern Indiana.

Douglas said, " We certainly cannot justify the holding of forts there ; much less the recapturing of those which have been taken, unless we intend to reduce those States themselves into subjection. We cannot deny that there is a Southern Confederacy, de facto, in existence. We may regret it. I regret it most profoundly, but I cannot deny the truth of the fact, painful and mortifying as it is."

Governor Price of New Jersey advised that State for business advantage to join the South. " To remain with the North separates us from those who have heretofore consumed our manufactures, and given employment to a large part of our labor ; our commerce will therefore cease ; our State becoming depopulated and impoverished, thereby affecting our agricultural interests. These are the prospective results of remaining with the present Northern Confederacy, whereas to join our destiny with the South will be to continue our trade and intercourse, perhaps in an augmented de-

gree.   Who is he that will advise New Jersey to pursue
the path of desolation, when one of prosperity is open
before us, without any sacrifice of principle or honor ?
. . .   The action of our State will prove influential
upon the adjoining great States of Pennsylvania and
New York.   It takes little discernment to see that such
a policy will enrich us and the other impoverish us.
The Constitution made at Montgomery has many modi-
fications desired by the people of this State.   We
believe that slavery is no sin ; that the negro is not
equal to the white man ; and that slavery is his normal
and natural condition."   Great mass meetings were
held in some of the Northern cities, nominally for peace.
For the most part these slumped into silly attempts to
placate the South with promises hereafter to obey the
Constitution.   Mayor Henry of Philadelphia presided
over a monster assemblage of a characteristic sort.
Among the resolutions were sentiments of this cast :
" Resolved that we recognize the obligations of the
Act of Congress of 1850, commonly known as the Fugitive Slave Law ; and submit cheerfully to its faithful
enforcement ; and that we point with pride and satisfaction to the recent conviction and punishment in
this city of Philadelphia of those who have broken its
provisions, by aiding in the attempted rescue of a slave,
as proof that Philadelphia is faithful in her obedience
to the law."   Not a few of the highest officers in the
army wavered for a time between loyalty to the Union
and fealty to their States.

But gradually the public sentiment of the loyal States
settled into a single conviction that this is not only a
compact of States, but is also a Nation.   Emphasis was
placed on unity ; and " The United States is " was
fondly substituted for " The United States are."   Both

statements are historically correct; but the former
needed more intelligent expression. That either statement alone is correct or safe is not true. Coming out
of the war there was not a smallest fraction anywhere
at the North, and none at the South worth the men-
tion, that did not accept the fact as inherent, that
peaceable secession is not a conceivable fact. Washington's idea of " an indissoluble Union " had become
triumphant and universal.

Dr. von Holst believes that the States, by accepting
the Constitution, fused thereby and at once into a nation ; and " so long as the people themselves, whose
work the Constitution is, do not decide to destroy the
Union, the Constitution of the Union despite any fact
whatever, remains from the standpoint of law un-
changed." This theory, if it means anything more
than the right of revolution, involves us in absurdity.
While denying the right of States to secede, it leaves
the people to break up the union when they will. This
must be by a majority vote ; and it follows that when
a majority decides to break the Union it is legitimately
severed. It follows further that had the secession of
1861 included more votes than remained in the Union,
the Union would really have been dissolved. That is,
the South could not legally secede in 1861 ; but the
North, by right of majority, could have done so. This
supposes a vast nation which may at any moment secede from a single State. The right to thus separate
in peace implies the right to separate in war. So negatived would be the end and purpose of union that no
pretense to unity would be worth the while. Such interpretation of the Constitution, by trying to conceive
a strongly organized nation, really ends in supposing a
vast people that at any moment may split apart by a

majority vote. The power of any State, or of any section to secede, we have tested in peace and settled in war. The federal idea must never be lost in that of the nation. The States must never be less integral than in 1788. The Federal Union of free and independent States can be broken only by successful revolution.

The progress of nationality has been a true evolution. Neither the States as factors of the Union nor the Union itself could remain at a standstill. Sovereignty held by the nation for a century as derived sovereignty has become a concrete right ; so that to-day it is doubtless true, as John Quincy Adams said, as long ago as 1830, that " the States hold from the Union as truly as the Union from the States." Brownson fairly sums up American history when he says, " The States without the Union cease to exist as political communities ; the Union without the States ceases to be a Union."

. . . " and becomes," he adds, " a vast centralized and consolidated State, ready to lapse from a civilized into a barbaric ; and from a republican to a despotic nation."

APPENDIX TO CHAPTER VII

COMPROMISE OF 1820

Be it further enacted that in all that Territory ceded by France to the United States, under the name of Louisiana, which lies north of 36° 30' north latitude, excepting only such part thereof as is included within the limits of the State contemplated by this act, slavery and involuntary servitude, otherwise than in the punishment of crime, whereof the party shall have been duly convicted, shall be and is hereby forever prohibited.   Provided always that any person escaping into the same, from whom labor or service is lawfully claimed in any State or Territory of the United States, such fugitive may be lawfully reclaimed, and conveyed to the person claiming his or her labor or service as aforesaid.

COMPROMISE OF 1850

Mr. Clay reported as chairman of the committee of thirteen the following bases of a general compromise, on May 8, 1850 :

1. The admission of any new State or States formed out of Texas to be postponed, until they shall hereafter present themselves to be received into the Union, when it will be the duty of Congress fairly and faithfully to execute the compact with Texas by admitting such new State or States.

2. The admission forthwith of California into the Union with the boundary which she has proposed.

3. The establishment of territorial government with-

out the Wilmot Proviso for New Mexico, and Utah, embracing all the territory recently acquired from Mexico not contained in the boundaries of California.

4. The combination of these two last measures in the same bill.

5. The establishment of western and northern boundaries of Texas, and the exclusion from her jurisdiction of all New Mexico, with the grant to Texas of a pecuniary equivalent, and the section for that purpose to be incorporated in the bill admitting California, and establishing Territorial government for Utah and New Mexico.

6. More effectual enactments of law to secure the prompt delivery of persons bound to service or labor in one State, under the laws thereof, who escape into another State.

7. Abstaining from abolishing slavery ; but under a heavy penalty prohibiting the slave trade, in the District of Columbia.

DIGEST OF THE OSTEND MANIFESTO

.   .   .   .

1. The United States ought, if practicable, to purchase Cuba, with as little delay as possible.

2. The probability is great that the Government and Cortes of Spain will prove willing to sell it, because this would essentially promote the highest and best interests of the Spanish people.

It must be clear to every reflecting mind that, from the peculiarity of its geographical position, and the considerations attendant on it, Cuba is as necessary to the North American republic as any of its present members, and that it belongs naturally to that great family of States of which the Union is the providential nursery.

From its locality it commands the mouth of the Mississippi and the immense and annually increasing trade which must seek this avenue to the ocean.

.   .   .   .

Our past history forbids that we should acquire the island of Cuba without the consent of Spain, unless justified by the great law of self-preservation. We must, in any event preserve our own conscious rectitude and our own self-respect.

While pursuing this course we can afford to disregard the censures of the world, to which we have been so often and so unjustly exposed.

After we shall have offered Spain a price for Cuba far beyond its present value, and this shall have been refused, it will then be time to consider the question, does Cuba, in the possession of Spain, seriously endanger our internal peace and the existence of our cherished Union.

Should this question be answered in the affirmative, then, by every law, human and divine, we shall be justified in wresting it from Spain if we possess the power ; and this upon the very same principle that would justify an individual in tearing down the burning house of his neighbor if there were no other means of preventing the flames from destroying his own home.

.   .   .   .

Yours, very respectfully,

JAMES BUCHANAN.
J. Y. MASON.
PIERRE SOULE.

ABOLITION OF THE COMPROMISE OF 1820,
MAY 30, 1854

Be it further enacted that in order to avoid all misconstruction it is hereby declared to be the true intent
and meaning of this act, so far as the question of slavery
is concerned, to carry into practical operation the following propositions and principles established by the Compromise measures of one thousand eight hundred and
fifty, to wit:

First, That all questions pertaining to slavery in the
Territories, and in the new States to be formed therefrom, are to be left to the decision of the people residing
therein, through their appropriate representatives.

Second, That all cases involving titles to slaves, and
questions of personal freedom, are referred to the adju-
dication of local tribunals ; with the right of appeal to
the Supreme Court of the United States.

Third, That the provisions of the Constitution and
laws of the United States in respect to fugitives from
service are to be carried into faithful execution in all
the organized Territories, the same as in the States.

Fourth, That the Constitution and all laws of the
United States, which are not locally inapplicable, shall
have the same force and effect within the said Territory
as elsewhere in the United States.

Except the section of the Act preparatory to the admission of Missouri into the Union, approved March 6,
1820, which was superceded by the principles of the
legislation of 1850, commonly called the Compromise
measure, and is declared inoperative.

JUDAH P. BENJAMIN ON SLAVES AS PROPERTY,
MARCH 11, 1858

Mr. President! the whole subject of slavery, so far as
it is involved in the issue now before the country, is
narrowed down at last to a controversy on the solitary
point whether it be competent for the Congress of the
United States, directly or indirectly, to exclude slavery
from the Territories of the Union. The Supreme Court
of the United States have given a negative answer to
this proposition ; and it shall be my first effort to support that negation by argument, independently of the
authority of the decision.

It seems to me that the radical, fundamental error,
which underlies the argument in affirmation of this
power, is the assumption that slavery is the creature of
the statute law of the several States where it is established; that it has no existence outside of the limits of
those States ; that slaves are not property beyond those
limits ; and that property in slaves is neither recognized
nor protected by the Constitution of the United States
nor by international law. I controvert all these propositions, and shall proceed at once to my argument.

Mr. President ! the thirteen colonies, which on the
4th of July, 1776, asserted their independence, were
British colonies, governed by British laws. Our ancestors, in their emigration to this country, brought with
them the common law of England, as their birthright.
They adopted its principles for their government, so
far as it was not incompatible with the peculiarities of
their situation in a rude and unsettled country. Great
Britain, then having the sovereignty over the colonies,
possessed undoubted power to regulate their institutions, to control their commerce, and to give laws to

their intercourse, both with the mother and the other
nations of the earth. If I can show, as I hope to be
able to establish to the satisfaction of the Senate, that
the nation thus exercising sovereign power over the
thirteen colonies did establish slavery in them, did
maintain and protect the institution, did originate and
carry on the slave trade, did support and foster that
trade ; that it forbade the colonies permission either to
emancipate or export their slaves, that it prohibited
them from inaugurating any legislation in diminution
or discouragement of the institution ; nay sir! more, if
at the date of our revolution I can show that African
slavery existed in England, as it did on this continent ;
if I can show that slaves were sold upon the slave
marts, in the Exchange, and other public places of resort in the city of London, as they were on this continent ; then I shall not hazard too much in the assertion
that slavery was the common law of the thirteen States
of the Confederacy, at the time they burst the bonds
that united them to the mother country. . . .

This legislation, Mr. President ! emanating from the
mother country, fixed the institution upon the colonies.
They could not resist it. All their right was limited to
petition, to remonstrance, and to attempts at legislation
at home to diminish the evil. Every such attempt was
sternly repressed by the British Crown. In 1760 South
Carolina passed an act prohibiting the further importation of African slaves. The act was rejected by the
Crown ; the governor was reprimanded ; and a circular
was sent to all the governors of all the colonies warning
them against presuming to countenance such legislation.
In 1774 two bills passed the Legislative Assembly of
Jamaica ; and the Earl of Dartmouth, then Secretary
of State, wrote to Sir Basil Keith the governor of the

colony, that " these measures had created alarm to the merchants of Great Britain engaged in that branch of commerce ; " and forbidding him " on pain of removal from his government to assent to such laws."

Finally in 1775—mark the date 1775—after the revolutionary struggle had commenced, whilst the Continental Congress was in session, after armies had been levied, after Crown Point and Ticonderoga had been taken possession of by the insurgent colonists, and after the first blood shed in the Revolution had reddened the spring sod upon the green at Lexington, this same Earl of Dartmouth, in remonstrance from the agent of the colonies, replied, " we cannot allow the colonies to check or discourage in any degree a traffic so beneficial to the nation."

But sir ! my task does not end here. I desire to show that by her jurisprudence, that by the decisions of her judges, and the answers of her lawyers, to questions from the Crown and from public bodies, this same institution was declared to be recognized by the common law of England ; and slaves were declared to be, in their language, " merchandise," chattels, just as much private property as any other merchandise or any other chattel. . . .

In the month of May, 1713, the British government undertook for the term of thirty years then next to come, to transport annually four thousand eight hundred slaves to the Spanish American colonies, at a fixed price. Almost immediately after this a question arose in the English Council as to what was the true legal character of the slaves thus to be exported to the Spanish American colonies ; and according to the forms of the British Constitution the question was submitted by the Crown in Council to the twelve judges of England.

I have their answer here. It is in these words : " In pursuance of his Majesty's order in Council, hereunto annexed, we do humbly certify our opinion to be that negroes are merchandise." . . .

Thus, Mr. President ! I say that, even if we admit for the moment that . . . the institution of slavery at the time of our Independence was dying away by the manumissions, either gratuitous or for a price, of those who held the people as slaves, yet so far as the continent of America was concerned, North and South, there did not breathe a being who did not know that a negro under the common law of the continent was merchandise, was property, was a slave; and that he could only extricate himself from that *status*, stamped upon him by the common law of the country, by positive proof of manumission. No man was bound to show title to his negro slaves. The slave was bound to show manumission, under which he had acquired his freedom by the common law of every colony. Why sir ! can any man doubt that if, after the revolution, the different States of the Union had not passed laws upon the subject, to abolish slavery, to subvert this common law of the continent, every one of these States would be slave States yet ? How came they free States ? Did not they have this institution of slavery imprinted upon them by the power of the mother country ? How did they get rid of it ? All must admit that they had to pass positive acts of legislation to accomplish this purpose. Without that legislation they would still be slave States. What, then becomes of the pretext that slavery only existed in those States where it was estabished by positive legislation ; that it has no inherent vitality out of those States ; and that slaves are not considered as property by the Constitution of the United States.

. . . Slaves, if you please, are not property like other property in this; that you can easily rob us of them; but as to the right in them, that man has to overthrow the whole history of the world, he has to overthrow every treatise on jurisprudence, he has to ignore the common sentiment of mankind, he has to repudiate the authority of all that is considered sacred with man, ere he can reach the conclusion that the person who owns a slave in a country where slavery has been established for ages, has no other property in that slave than the mere title which is given by the statute law of the land where it is found. . . .

SOUTH CAROLINA'S ORDINANCE OF SECESSION, DEC. 20, 1860

An Ordinance

To dissolve the Union between the State of South Carolina and other States united with her under the compact entitled "The Constitution of the United States of America."

We, the People of the State of South Carolina, in Convention assembled, do declare and ordain, and it is hereby declared and ordained.

That the Ordinance adopted by us in Convention, on the twenty-third day of May, in the year of our Lord one thousand seven hundred and eighty-eight, whereby the Constitution of the United States of America was ratified, and also, all Acts and parts of Acts of the General Assembly of this State, ratifying amendments of the said Constitution, are hereby repealed; and that the Union now subsisting between South Carolina and other States, under the name of "The United States of America," is hereby dissolved.

25

SOUTH CAROLINA'S DECLARATION OF CAUSES WHICH
    INDUCED HER SECESSION FROM THE FEDERAL
    UNION, DEC. 24, 1860

Declaration of the immediate causes which induce
and justify the secession of South Carolina from the
Federal Union.

The People of the State of South Carolina, in Convention assembled, on the 26th day of April, A.D., 1852,
declared that the frequent violation of the Constitution
of the United States, by the Federal Government, and
its encroachments upon the reserved rights of the
States, fully justified this State in then withdrawing
from the Federal Union ; but in deference to the opinions and wishes of the other slaveholding States, she
forebore at that time to exercise this right.   Since that
time, these encroachments have continued to increase,
and further forbearance ceases to be a virtue.

And now the State of South Carolina having resumed
her separate and equal place among nations, deems it
due to herself, to the remaining United States of America, and to the nations of the world, that she should declare the immediate causes which have led to this act.

In the year 1765, that portion of the British Empire
embracing Great Britain, undertook to make laws for
the government of that portion composed of the thirteen
American Colonies.   A struggle for the right of self-
government ensued, which resulted, on the 4th of July,
1776, in a Declaration, by the Colonies, " that they are
and of right ought to be, Free and Independent States ;
and that, as free and independent States, they have full
power to levy war, conclude peace, contract alliances,
establish commerce, and to do all other acts and things
which independent States may of right do.''

They further solemnly declared that whenever any "form of government becomes destructive of the ends for which it was established, it is the right of the people to alter or abolish it, and to institute a new government." Deeming the government of Great Britain to have become destructive of these ends, they declared that the Colonies "are absolved from all allegiance to the British Crown, and that all political connection between them and the State of Great Britain is, and ought to be totally dissolved."

In pursuance of this Declaration of Independence, each of the thirteen States proceeded to exercise its separate sovereignty; adopted for itself a Constitution, and appointed officers for the administration of government in all its departments—Legislative, Executive, and Judicial. For purposes of defence, they united their arms and their counsels: and, in 1778, they entered into a League known as the Articles of Confederation, whereby they agreed to entrust the administration of their external relations to a common agent, known as the Congress of the United States; expressly declaring in the first article, " that each State retains its sovereignty, freedom, and independence, and every power, jurisdiction and right which is not, by this Confederation, expressly delegated to the United States in Congress assembled."

Under this Confederation the War of the Revolution was carried on, and on the 3d September, 1783, the contest ended, and a definite Treaty was signed by Great Britain, in which she acknowledged the Independence of the Colonies in the following terms.

" Article 1.—His Britannic Majesty acknowledges the said United States, viz : New Hampshire, Massachusetts Bay, Rhode Island and Providence Plantations, Connecticut, New York, New Jersey, Pennsylvania, Delaware, Maryland, Virginia, North Carolina, South Carolina, and Georgia, to be Free, Sovereign, and Independent States ; that he treats with them as such ; and for himself, his heirs and successors, relinquishes all claims to the government, property and territorial rights of the same and every part thereof.''

Thus were established the two great principles asserted by the Colonies, namely : the right of a State to govern itself; and the right of a people to abolish a Government when it becomes destructive of the ends for which it was instituted. And concurrent with the establishment of these principles, was the fact, that each Colony became and was recognized by the mother Country as a Free, Sovereign, and Independent State.

In 1787, Deputies were appointed by the States to revise the Articles of Confederation, and on 17th September, 1787, these Deputies recommended for the adoption of the States, the Articles of Union, known as the Constitution of the United States.

The parties to whom this Constitution was submitted, were the several sovereign States ; they were to agree or disagree, and when nine of them agreed, the compact was to take effect among those concurring ; and the General Government, as the common agent, was then to be invested with their authority.

If only nine of the thirteen States had concurred, the other four would have remained as they were—separate sovereign States, independent of any of the provisions of the Constitution. In fact, two of the States did not accede to the Constitution until long after it had gone into operation among the other eleven, and during that interval, they each exercised the functions of an independent nation.

By this Constitution, certain duties were imposed upon the several States, and the exercise of certain of their powers was restrained, which necessarily implied their continued existence as sovereign States. But, to remove all doubt, an amendment was added, which declared that the powers not delegated to the United States by the Constitution, nor prohibited by it to the States, are reserved to the States, respectively, or to the people. On 23d May, 1788, South Carolina, by a Convention of her people, passed an Ordinance assenting to this Constitution, and afterwards altered her own Constitution, to conform herself to the obligations she had undertaken.

Thus was established, by compact between the States, a Government, with defined objects and powers, limited to the express words of the grant. This limitation left the whole remaining mass of power subject to the clause reserving it to the States or to the people, and rendered unnecessary any specification of reserved rights.

We hold that the Government thus established is subject to the two great principles asserted in the Declaration of Independence ; and we hold further, that the mode of its formation subjects it to a third fundamental principle, namely : the law of compact. We maintain that in every compact between two or more parties, the obligation is mutual; that the failure of one of the contracting parties to perform a material part of the agreement, entirely releases the obligation of the other ; and that where no arbiter is provided, each party is remitted to his own judgment to determine the fact of failure, with all its consequences.

In the present case, that fact is established with certainty. We assert, that fourteen of the States have

deliberately refused for years past to fulfil their constitutional obligations, and we refer to their own Statutes
for the proof.

The Constitution of the United States, in its 4th
Article provides as follows :

" No person held to service or labor in one State,
under the laws thereof, escaping into another, shall, in
consequence of any law or regulation therein, be discharged from such service or labor, but shall be delivered up, on claim of the party to whom such service
may be due."

This stipulation was so material to the compact, that
without it that compact would not have been made. The
greater number of the contracting parties held slaves,
and they had previously evinced their estimate of the
value of such a stipulation by making it a condition in
the Ordinance for the government of the territory ceded
by Virginia, which now composes the States north of
the Ohio river.

The same article of the Constitution stipulates also
for rendition by the several States of fugitives from
justice from the other States.

The General Government, as the common agent,
passed laws to carry into effect these stipulations of the
States. For many years these laws were executed.
But an increasing hostility on the part of the non-
slaveholding States to the Institution of Slavery, has
led to a disregard of their obligations, and the laws
of the General Government have ceased to effect the
objects of the Constitution. The States of Maine, New
Hampshire, Vermont, Massachusetts, Connecticut,
Rhode Island, New York, Pennsylvania, Illinois, Indiana, Michigan, Wisconsin, and Iowa, have enacted
laws which either nullify the Acts of Congress or ren-

der useless any attempt to execute them. In many of
these States the fugitive is discharged from the service
or labor claimed, and in none of them has the State
Government complied with the stipulation made in the
Constitution. The State of New Jersey, at an early
day, passed a law in conformity with her constitutional
obligation ; but the current of anti-slavery feeling has
led her more recently to enact laws which render inoperative the remedies provided by her own law and by
the laws of Congress. In the State of New York even
the right of transit for a slave has been denied by her
tribunals ; but the States of Ohio and Iowa have refused to surrender to justice fugitives charged with
murder, and with inciting servile insurrection in the
State of Virginia. Thus the constitutional compact
has been deliberately broken and disregarded by the
non-slaveholding States, and the consequence follows
that South Carolina is released from her obligation.

The ends for which this Constitution was framed are
declared by itself to be " to form a more perfect union,
establish justice, insure domestic tranquility, provide
for the common defence, promote the general welfare,
and secure the blessings of liberty to ourselves and our
posterity."

These ends it endeavored to accomplish by a Federal
Government, in which each State was recognized as
an equal, and had separate control over its own institutions. The right of property in slaves was recognized, by giving to free persons distinct political rights;
by giving them the right to represent, and burthening
them with direct taxes for three-fifths of their slaves ;
by authorizing the importation of slaves for twenty
years; and by stipulating for the rendition of fugitives
from labor.

We affirm that these ends for which this Government was instituted have been defeated, and the Government itself has been made destructive of them by the action of the non-slaveholding States. Those States have assumed the right of deciding upon the propriety of our domestic institutions ; and have denied the rights of property established in fifteen of the States, and recognized by the Constitution ; they have denounced as sinful the institution of Slavery; they have permitted the open establishment among them of societies, whose avowed object is to disturb the peace and to eloign the property of the citizens of other States ; they have encouraged and assisted thousands of our slaves to leave their homes ; and those who remain, have been incited by emissaries, books and pictures to servile insurrection.

For twenty-five years this agitation has been steadily increasing, until it has now secured to its aid the power of the Common Government. Observing the forms of the Constitution, a sectional party has found within that article establishing the Executive Department, the means of subverting the Constitution itself. A geographical line has been drawn across the Union, and all the States north of that line have united in the election of a man to the high office of President of the United States whose opinions and purposes are hostile to slavery. He is to be entrusted with the administration of the Common Government, because he has declared that that " Government cannot endure permanently half slave, half free," and that the public mind must rest in the belief that Slavery is in the course of ultimate extinction.

This sectional combination for the subversion of the Constitution, has been aided in some of the States by elevating to citizenship, persons, who, by the Supreme Law of the land, are incapable of becoming citizens ; and their votes have been used to inaugurate a new policy, hostile to the South, and destructive of its peace and safety.

On the 4th March next, this party will take possession of the Government. It has announced, that the South shall be excluded from the common territory ; that the Judicial Tribunals shall be made sectional, and that a war must be waged against slavery until it shall cease throughout the United States.

The Guarantees of the Constitution will then no longer exist ; the equal rights of the States will be lost. The slaveholding States will no longer have the power of self-government, or self-protection, and the Federal Government will have become their enemy.

Sectional interest and animosity will deepen the irritation, and all hope of remedy is rendered vain, by the fact that public opinion at the North has invested a great political error with the sanctions of a more erroneous religious belief.

We, therefore, the people of South Carolina, by our delegates in Convention assembled, appealing to the Supreme Judge of the world for the rectitude of our intentions, have solemnly declared that the Union heretofore existing between this State and the other States of North America, is dissolved ; and that the State of South Carolina has resumed her position among the nations of the world, as a separate and independent State ; with full power to levy war, conclude peace, contract alliances, establish commerce, and to do all other acts and things which independent States may of right do.

ALEXANDER HAMILTON STEPHENS CONCERNING THE
CONSTITUTION OF THE CONFEDERACY, MARCH
21, 1861

All the essentials of the old Constitution, which have
endeared it to the hearts of the American people, have
been preserved and perpetuated. Some changes have
been made. Some of these I should prefer not to
have been made ; but other important changes meet my
cordial approbation. . . . Allow me briefly to allude
to some of these improvements. The question of building up class interests or fostering one branch of industry to the prejudice of another, under the exercise of
the revenue power, which gave us so much trouble
under the old Constitution, is put at rest forever under
the new. We allow the imposition of no duty, with
the view of giving an advantage to one class of persons
in any trade or business over those of another. All,
under our system, stand upon the same broad principles
of perfect equality. Honest labor and enterprise are
left free and unrestricted, in whatever pursuit they may
be engaged. This old thorn of the tariff, which was
the cause of so much irritation in the old body politic,
is removed forever from the new.

Again the subject of internal improvements, under
the power of Congress to regulate commerce, is put at
rest under our system. The power claimed, by construction, under the old Constitution, was at least a
doubtful one ; it rested solely upon construction. We
of the South generally, apart from considerations of
constitutional principles, opposed its exercise upon
grounds of its inexpediency and injustice. . . . No
State was in greater need of such facilities than Georgia,
but we did not ask that these works should be made by
appropriations out of the common treasury. The cost

of the grading, the superstructure, and the equipment
of our roads was borne by those who had entered into
the enterprise. Nay, more, not only the cost of the
iron was borne in the same way, but we were compelled
to pay into the common treasury several millions of
dollars, for the privilege of importing the iron, after the
price was paid for it abroad. What justice was there
in taking this money, which our people paid into the
common treasury, on the importation of our iron, and
applying it to the improvement of rivers and harbors
elsewhere ? The true principle is to subject the commerce of every locality to whatever burdens may be
necessary to facilitate it.   .   .   .

Another feature to which I will allude is that the
new Constitution provides that Cabinet ministers and
Heads of Departments may have the privilege of seats
upon the floor of the Senate, and House of Representatives ; may have the right to participate in the debates
and discussions upon the various subjects of administration. I should have preferred that this provision
should have gone further, and required the President
to select his Constitutional advisers from the Senate
and House of Representatives. That would have conformed entirely to the practice in the British Parliament ; which in my judgment is one of the wisest
provisions in the British Constitution.   .   .   .

Another change in the Constitution relates to the
length of the tenure of the Presidential office. In the
new Constitution it is six years instead of four ; and the
President is rendered ineligible for a re-election. This
is certainly a decidedly conservative change. It will
remove from the incumbent all temptation to use his
office or exert the powers confided to him for any object
of personal ambition.

But not to be tedious in enumerating the numerous changes for the better, allow me to allude to one other, the last, not least. The new Constitution has put at rest forever all the agitating questions relating to our peculiar institution, African Slavery. This was the immediate cause of the late rupture and present revolution. Jefferson, in his forecast, had anticipated this as the "rock upon which the old Union would split." He was right. What was conjecture with him is now a realized fact. But whether he fully comprehended the great truth upon which that rock stood and stands may be doubted. The prevailing ideas entertained by him, and most of the leading statesmen at the time of the formation of the old Constitution, were that the enslavement of the African was in violation of the laws of nature, that it was wrong in principle, socially, morally, and politically. It was an evil they knew not well how to deal with ; but the general opinion of the men of that day was that, somehow or other, in the order of Providence, the institution would be evanescent and pass away. This idea, though not incorporated in the Constitution, was the prevailing idea at that time. The Constitution, it is true, secured every essential guarantee to the institution while it should last, and hence no argument can be justly urged against the Constitutional guarantees thus secured, because of the common sentiment of the day. Those ideas, however, were fundamentally wrong. They rested upon the assumption of the equality of races. This was an error. It was a sandy foundation, and the government built upon it fell "when the storm came and the wind blew." Our new government is founded upon exactly the opposite idea. Its foundations are laid, its cornerstone rests upon the great truth that the negro is not

equal to the white man ; that slavery—subordination
to the superior race—is his natural and normal condition.   This, our new government, is the first in the
history of the world based upon this great physical,
philosophical, and moral truth.   .   .   .

Many governments have been founded upon the
principle of the subordination and serfdom of certain
classes of the same race ; such were and are in violation
of the laws of nature.   Our system commits no such
violation of nature's laws.   With us all the white race,
however high or low, rich or poor, are equal in the eye
of the law.   Not so with the negro ; subordination is
his place.

The substratum of our society is made of the material
fitted by nature for it ; and by experience we know that
it is best not only for the superior race, but for the inferior race that it should be so.   .   .   .   It is indeed
in conformity with the ordinance of the Creator :   It is
not for us to inquire into the wisdom of his ordinances
or to question them.   The great objects of humanity
are best attained when there is conformity to His laws
and decrees, in the formation of governments as well as
in all things else.   Our Confederacy is founded upon
principles in strict conformity with these views.   This
stone which was rejected by the first builders is become
" the chief of the corner," the real corner-stone in our
new edifice.

JEFFERSON DAVIS'S VIEW OF THE RIGHT TO SECEDE.
     — ON HIS WITHDRAWAL FROM THE SENATE,
     JANUARY 21, 1861

Nullification and secession so often confounded are
antagonistic principles.   Nullification is a remedy

which it is sought to apply within the Union, and against the agent of the States. It is only to be justified when the agent has violated his constitutional obligation ; and a State assuming to judge for itself, denies the right of the agent thus to act, and appeals to the other States of the Union for a decision ; but when the States themselves, and when the people of the States have so acted as to convince us that they will not regard our Constitutional rights, then, and then for the first time, arises the doctrine of secession in its practical application.

A great man who now reposes with his fathers, and who has been often arraigned for a want of fealty to the Union, advocated the doctrine of nullification because it preserved the Union. It was because of his deepseated attachment to the Union, his determination to find some remedy for existing ills short of a severance of the ties which bound South Carolina to the other States, that Mr. Calhoun advocated the doctrine of nullification, which he proclaimed to be peaceful, and to be within the limits of State power ; not to disturb the Union, but only to be a means of bringing the agent before the tribunal of the States, for their judgment.

Secession belongs to a different class of remedies. It is to be justified upon the basis that the States are sovereign. There was a time when none denied it. I hope the time may come again when a better comprehension of the theory of our Government, and the inalienable rights of the people of the States, will prevent any one from denying that each State is a sovereign ; and thus may reclaim the grants which it has made to any agent whomsoever.

ARGUMENT OF ALFRED IVERSON CONCERNING THE
RIGHT TO SECEDE, DECEMBER, 1860

I do not myself place the right of a State to secede
from the Union upon Constitutional grounds. I admit
that the Constitution has not granted that power to a
State. It is exceedingly doubtful even whether the
right has been reserved. Certainly it has not been reserved in express terms. I therefore do not place the
expected action of any of the Southern States in the
present contingency, upon the constitutional right of
secession ; and I am not prepared to dispute therefore
the position which the President has taken upon that
point.

I rather agree with the President that the secession
of a State is an act of revolution ; taken through that
particular means or by that particular measure. It
withdraws from the Federal compact, disclaims any
further allegiance to it, and sets itself up as a separate
government, an independent State. The State does it
at its peril, of course ; because it may, or may not, be
cause of war by the remaining States composing the
Federal Government. If they think proper to consider
it such an act of disobedience, or if they consider that
the policy of the Federal Government be such that it
cannot submit to this dismemberment, why then they
may or may not make war, as they choose, upon the
seceding States. It will be a question of course for the
Federal Government, or the remaining States, to decide
for themselves, whether they will permit a State to go
out of the Union, and remain as a separate and independent State, or whether they will attempt to force
her back at the point of the bayonet. That is a question, I presume, of policy and expediency, which will

be considered by the remaining States composing the Federal Government, through their organ, the Federal Government, whenever the contingency arises.

But sir ! while a State has no power under the Constitution conferred upon it, to secede from the Federal Government or from the Union, each State has the right of revolution, which all admit. Whenever the burdens of the Government under which it acts become so onerous that it cannot bear them, or if anticipated evil shall be so great that the State believes it would be better off—even risking the perils of secession—out of the union than in it, then that State, in my opinion, like all people upon earth, has the right to exercise the great fundamental principle of self-preservation, and go out of the Union—though of course at its own peril. And while no State may have the constitutional right to secede from the Union, the President may not be wrong when he says the Federal Government has no power under the Constitution to compel the State to come back into the Union. It may be a *casus omissus* in the Constitution ; but I should like to know where the power exists in the Constitution of the United States to authorize the Federal Government to coerce a sovereign State. It does not exist in terms, at any rate, in the Constitution. . . .

GREELEY ON THE RIGHT OF SECESSION,
FEBRUARY, 1861

The telegraph informs us that most of the Cotton States are meditating a withdrawal from the Union, because of Lincoln's election. Very well: they have a right to meditate, and meditation is a profitable employment of leisure. We have a chronic, invincible disbe-

lief in Disunion as a remedy for either Northern or
Southern grievances. We cannot see any necessary
connection between the alleged disease and this ultra-
heroic remedy; still, we say, if any one sees fit to meditate Disunion, let him do so unmolested. That was a
base and hypocritic row that was once raised, at Southern dictation, about the ears of John Quincy Adams,
because he presented a petition for the dissolution of
the Union. The petitioner had a right to make the
request ; it was the Member's duty to present it. And
now, if the Cotton States consider the value of the
Union debatable, we maintain their perfect right to discuss it. Nay : we hold, with Jefferson, to the inalienable right of communities to alter or abolish forms of
government that have become oppressive or injurious ;
and, if the Cotton States shall decide that they can do
better out of the Union than in it, we insist on letting
them go in peace. The right to secede may be a revolutionary one, but it exists nevertheless ; and we do
not see how one party can have a right to do what another party has a right to prevent. We must ever
resist the asserted right of any State to remain in the
Union, and nullify or defy the laws thereof ; to withdraw from the Union is quite another matter. And,
whenever a considerable section of our Union shall deliberately resolve to go out, we shall resist all coercive
measures designed to keep it in. We hope never to
live in a republic, whereof one section is pinned to the
residue by bayonets.

But, while we thus uphold the practical liberty, if not
the abstract right, of secession, we must insist that the
step be taken, if it ever shall be, with the deliberation
and gravity befitting so momentous an issue. Let
ample time be given for reflection ; let the subject be

26

fully canvassed before the people ; and let a popular vote be taken in every case, before Secession is decreed. Let the people be told just why they are asked to break up the confederation ; let them have both sides of the question fully presented ; let them reflect, deliberate, then vote ; and let the act of Secession be the echo of an unmistakable popular fiat. A judgment thus rendered, a demand for separation so backed, would either be acquiesced in without the effusion of blood; or those who rushed upon carnage to defy and defeat it, would place themselves clearly in the wrong.

ABRAHAM LINCOLN'S VIEW OF THE RIGHT TO SECEDE.—FROM HIS FIRST INAUGURAL, MARCH 4, 1861

I hold that in contemplation of universal law, and of the Constitution, the union of these States is perpetual. Perpetuity is implied if not expressed in the fundamental law of all national governments. It is safe to assert that no government proper ever had a provision in its organic law for its own termination. Continue to execute all the expressed provisions of our National Government, and the Union will endure forever—it being impossible to destroy it except by some action not provided for in the instrument itself.

Again if the United States be not a government proper, but an association of States, in the nature of contract merely, can it as a contract be peaceably unmade by less than all the parties who made it ? One party to a contract may violate it,—break it so to speak ; but does it not require all to lawfully rescind it ?

Descending from these general principles, we find the proposition, that in legal contemplation the Union is

perpetual, confirmed by the history of the Union itself.
The Union is much older than the Constitution. It
was formed in fact by the Articles of Association in
1774. It was matured and continued by the Declaration of Independence in 1776. It was further matured,
and the faith of all the then thirteen States expressly
plighted and engaged that it should be perpetual, by
the Articles of Confederation in 1778. And finally in
1787 one of the declared objects for ordaining and establishing the Constitution was to form a more perfect
union.

But if destruction of the Union by one, or by a part
only of the States, be lawfully possible, the Union is
less perfect than before ; the Constitution having lost
the vital element of perpetuity.

It follows from these views, that no State, upon its
own mere motion, can lawfully get out of the Union ;
that resolves and ordinances to that effect are legally
void ; and that acts of violence within any State or
States against the authority of the United States, are
insurrectionary, or revolutionary, according to circumstances.

I therefore consider that in view of the Constitution
and the laws the Union is unbroken ; and to the extent
of my ability I shall take care as the Constitution itself
expressly enjoins upon me, that the laws of the Union
be faithfully executed in all the States.

.   .   .   If a minority in such case will secede rather
than acquiesce, they make a precedent, which in turn
will divide and ruin them ; for a minority of their own
will secede from them whenever a majority refuses to
be controlled by a minority.   .   .   .

Plainly the central idea of secession is the essence of
anarchy. A majority held in restraint by Constitu-

tional checks and limitations, and always changing
easily with deliberate changes of popular opinions and
sentiments, is the only true sovereign of a free people.
Whoever rejects it, does of necessity fly to anarchy or
to despotism. . . . The Chief Magistrate derives
all his authority from the people and they have conferred none upon him to fix terms for the separation of
the States. His duty is to administer the present government as it came to his hands, and to transmit it unimpaired by him to his successor. Why should there
not be a patient confidence in the ultimate justice of the
people ? Is there any better or equal hope in the world?
In our present differences is either party without faith
of being in the right ? By the frame of the government
under which we live, the same people have wisely given
their public servants but little power for mischief, and
have with equal wisdom provided for the return of that
little to their own hands at very short intervals. While
the people retain their virtue and vigilance, no administration by any extreme of wickedness or folly, can
very seriously injure the government in the short space
of four years.

. . . In your hands, my dissatisfied fellow-
countrymen, and not in mine, are the momentous issues
of civil war. The government will not assail you.
You can have no conflict without being yourselves the
aggressors. You have no oath registered in heaven to
destroy the government, while I shall have the most
solemn one to preserve, protect, and defend it.

PROCLAMATION OF EMANCIPATION

Whereas, on the 22d day of September, in the year
of our Lord 1862, a proclamation was issued by the

President of the United States, containing, among other things, the following, to wit :

" That on the 1st day of January, in the year of our Lord 1863, all persons held as slaves within any State or designated part of a State, the people whereof shall then be in rebellion against the United States, shall be then, thenceforward, and forever free ; and the Executive Government of the United States, including the military and naval authority thereof, will recognize and maintain the freedom of such persons, and will do no act or acts to repress such persons, or any of them, in any efforts they may make for their actual freedom.

" That the Executive will, on the first day of January aforesaid, by proclamation, designate the States and parts of States, if any, in which the people thereof respectively shall then be in rebellion against the United States ; and the fact that any State, or the people thereof, shall on that day be in good faith represented in the Congress of the United States, by members chosen thereto at elections wherein a majority of the qualified voters of such State shall have participated, shall, in the absence of strong countervailing testimony, be deemed conclusive evidence that such State, and the people thereof, are not then in rebellion against the United States."

Now therefore, I, Abraham Lincoln, President of the United States, by virtue of the power in me vested as Commander-in-Chief of the Army and Navy of the United States, in time of actual armed rebellion against the authority and Government of the United States, and as a fit and necessary war measure for suppressing said rebellion, do, on this first day of January, in the year of our Lord one thousand eight hundred and sixty-three, and in accordance with my purpose so to do, publicly proclaimed for the full period of one hundred days from the first day above mentioned, order and designate as the States and parts of States wherein the people thereof respectively are this day in rebellion against the United States, the following, to wit :

Arkansas, Texas, Louisiana (except the parishes of St. Bernard Plaquemine, Jefferson, St. John, St. Charles, St. James, Ascension, Assumption, Terre Bonne, Lafourche, St. Mary, St. Martin, and Orleans, including the city of New Orleans), Mississippi, Alabama, Florida, Georgia, South Carolina, North Carolina, and Virginia (except the forty-eight counties designated as West Virginia, and also the counties of Berkeley, Accomac, Northampton, Elizabeth City, York, Princess Anne, and Norfolk, including the cities of Norfolk and Portsmouth), and which excepted parts are, for the present, left precisely as if this proclamation were not issued.

And, by virtue of the power and for the purpose aforesaid, I do order and declare that all persons held as slaves within said designated States and parts of States, are, and henceforward shall be free ; and that the Executive Government of the United States, including the military and naval authorities thereof, will recognize and maintain the freedom of said persons.

And I hereby enjoin upon the people so declared to be free to abstain from all violence, unless in necessary self-defense ; and I recommend to them that, in all cases when allowed, they labor faithfully for reasonable wages.

And I further declare and make known that such persons, of suitable condition, will be received into the armed service of the United States to garrison forts, positions, stations, and other places, and to man vessels of all sorts in said service.

And upon this act, sincerely believed to be an act of justice, warranted by the Constitution upon military necessity, I invoke the considerate judgment of mankind, and the gracious favor of Almighty God.

In testimony whereof, I have hereunto set my name, and caused the seal of the United States to be affixed.

Done at the city of Washington, this 1st day of January, in the year of our Lord 1863, and of the independence of the United States the 87th.

By the President : ABRAHAM LINCOLN.
(L.S.)

WILLIAM H. SEWARD, Secretary of State.

AMENDMENTS TO THE CONSTITUTION FOLLOWING EMANCIPATION

Article XIII, Ratified Dec. 18, 1865

Section 1. Neither slavery nor involuntary servitude, except as a punishment for crime, whereof the party shall have been duly convicted, shall exist within the United States, or any place subject to their jurisdiction.

Sec. 2. Congress shall have power to enforce this article by appropriate legislation.

Instead of this Amendment the Crittenden Compromise of Dec. 18, 1860, had proposed,

Article XIII.

Section 1. Slavery shall be prohibited in all the territories of the United States now held or hereafter acquired situate north of latitude 36° 30'. In all territories south of said line of latitude slavery is hereby recognized as existing, and shall not be interfered with

by Congress ; but shall be protected as property by all
the departments of the territorial government during
its continuance.

Sec. 2. Congress shall have no power to abolish
slavery in places under its exclusive jurisdiction, and
situate within the limits of States that permit the holding of slaves.

Sec. 3. Congress shall have no power to abolish
slavery in the District of Columbia without compensation—,and without the consent of its inhabitants, and
of Virginia and of Maryland.

Sec. 4. Congress shall have no power to prohibit or
hinder the transportation of slaves between slaveholding States and territories.

Sec. 5. (A provision for the payment of the owners
by the United States for rescued fugitive slaves.)

Sec. 6. No future amendment of the Constitution
shall affect the five preceding sections ; and no amendment shall be made to the Constitution which will
authorize or give to Congress any power to abolish or
interfere with slavery in any of the States by whose
laws it is or may be allowed or permitted.

Article XIV, Ratified July 28, 1868

Section 1. All persons naturalized in the United
States, and subject to the jurisdiction thereof, are citizens of the United States and of the State wherein they
reside. No State shall make or enforce any law which
shall abridge the privileges or immunities of citizens of
the United States, nor shall any State deprive any person of life, liberty, or property without due process of
law, nor deny to any person within its jurisdiction the
equal protection of the laws.

Sec. 2. Representatives shall be apportioned among the several States according to their respective numbers, counting the whole number of persons in each State, excluding Indians not taxed. But when the right to vote at any election for the choice of electors for President and Vice President of the United States, representatives in Congress, the executive and judicial officers of a State, or the members of the legislature thereof, is denied to any of the male inhabitants of such State, being twenty-one years of age, and citizens of the United States ; or in any way abridged, except for participation in rebellion or other crime, the basis of representation therein shall be reduced in the proportion which the number of such male citizens shall bear to the whole number of male citizens twenty-one years of age in such State.

Sec. 3. No person shall be a Senator or Representative in Congress, or elector of President and Vice-President, or hold any office, civil or military, under the United States, or under any State, who, having previously taken an oath as a member of Congress, or as an officer of the United States, or as a member of any State legislature, or as an executive or judicial officer of any State, to support the Constitution of the United States, shall have engaged in insurrection, or rebellion against the same, or given aid or comfort to the enemies thereof. But Congress may, by a vote of two-thirds of each House, remove such disability.

Sec. 4. The validity of the public debt of the United States authorized by law, including debts incurred for payment of pensions and bounties for services in suppressing insurrection or rebellion, shall not be questioned. But neither the United States nor any State shall assume or pay any debt or obligation incurred in aid of insurrection or rebellion against the United States, or any claim for the loss or emancipation of any slave ; but all such debts, obligations, and claims shall be held illegal and void.

Sec. 5. The Congress shall have power to enforce, by appropriate legislation, the provisions of this article.

Article XV, Ratified March 30, 1870

Section 1. The right of citizens of the United States to vote shall not be denied or abridged by the United States, or by any State, on account of race, color, or previous condition of servitude.

Sec. 2. The Congress shall have power to enforce this article by appropriate legislation.

CONCLUDING CHAPTER

MASSACHUSETTS was founded by companies of people who left England because they had broken with Church, or with State, or with both. For fifty years they governed themselves as they chose. Other colonies had a similar experience. When the mother country began to assert supremacy it was too late to establish it. Tossing the tea into Boston Harbor was nullification of the Stamp Act; the Declaration of Independence consummated a scission, which had practically existed from the outset. Meanwhile there had been twenty-three plans of Union devised between the Colonies; all proving ineffectual, because not one of them yielded the ancient right of a State to tax the commerce of its neighbors. Yet all plans pointed toward, and gradually led to the Constitution of 1787, which abolished and forbade impost taxes between the States. John Adams says that "The Constitution was wrung from the necessities of a reluctant people." The marvel is that no more than six attempts at a breach of fellowship occurred within the first century of national existence.

The questions involved in the attempts at nullification and secession have so far been, The right of States to decide the Constitutionality of Congressional Acts; The right of the select few to control the politics of the

people ; The right of genius to supremacy ; The right
to resist laws that bear unequally on sections ; The
possibility of holding in union antagonistic social
forces. All these questions have been solved favorably
to the Republic. But are we sure that other questions
may not arise, or are not now under discussion, which
will disturb the political solidarity ?

First.—Referring to an attempt in 1779 to enact
emancipation in Virginia, Jefferson said, " It was
found that the public mind would not bear the proposition—nor will it to this day. Yet the day is not distant when it must bear it, and adopt it, or worse will
follow." But to set free four millions of slaves, and
endow them with the privileges of citizenship, did not
bring us to the end of trouble with the Negro Problem.
Optimism and philanthropy cannot blind us to the
complex evils of race antagonism. This is certain that
the blacks will not at present be allowed to exercise
their Constitutional right of suffrage. For some time
after the War they were voted by carpet-bag Northern-
ers, they are voted now by native-born Southerners.
The remedy offered by education, although probably the
only one in which the nation can place reliance, is not
immediate. Booker Washington has succeeded in de-
monstrating that this education must add industrial
training to intellectual culture.

Apart from the question of suffrage, the South at
present offers no hindrance to federalism. Railroads run
North and South as freely as East and West, creating
a community of population and of popular sentiment.
The literature of the South, freed from the influence of
the " irrepressible sentiment," is competing fairly with
that of the North. Churches that split across the sectional line, are either re-uniting, or fusing in the larger

liberty of a religion of humanity. The self-righteousness of the North, consequent upon a long struggle for a moral principle, is fading no faster than the self-satisfaction of the South with patriarchal society and mediæval customs. We can dismiss the era of sectionalism with confident assurance, that the long rivalry of North and South, of free and slave labor, was of inestimable benefit in developing strength of national character, and a completer knowledge of what our institutions are good for. Our fathers, enumerating the checks established by the Constitution on possible usurpation, did not foresee that but for the influence of one section upon another, in our first century, liberty would have been sacrificed to government. The factional issue went into history with ultimate profit to our national life ; so also did the sectional contest.

Second.—The same problem of how to deal with ignorance in a government of equal suffrage, confronts us elsewhere as well as with the negro. The firm conviction expressed by Washington, Jefferson, and others among the nation builders, that a republic of freemen could cohere only as cemented by popular intelligence and general education, was not novel to them. It was borrowed of Bradford, Winslow, and Roger Williams. It became an inheritance of the whole nation. If, of late, belief in mere education is expressed with a less confident tone, it is because there is a growing conviction that educational system and substance must improve as rapidly as political theory and social manners. The world heretofore has never seen such universal zeal among educators as is now manifested in the United States. Defective in civic instruction, and in moral stimulus, as our common school system has been, it cannot long remain so. Mediævalism in our higher

institutions is also steadily giving way to the science
and spirit of the present era. From Horace Mann to
W. T. Harris is but a generation, but it marks a stride
from local feebleness to organic strength. However,
the work ahead is greater than that already accomplished.

Our great cities have proved victims and tools of
corruption. Bismarck flung at us, " You Americans
cannot manage great municipalities." Carlyle sneered
that, " Judas is equal to Jesus in your system of democratic balloting." But we cannot forget that the drift
toward concentration of population in herded cities has
been largely due to the development of steam power ;
and now we are assured that the Steam Age is nearing
its end. Mr. Orton, our best American authority, tells
us that our coal deposits will be practically exhausted
by the middle of the twentieth century. A Parliamentary Commission reports that English coal cannot out-
last one hundred years. The Steam Age began about
1830 ; it will not run beyond 1930. Steam power has
been a concentering force ; electricity, on the contrary,
can be carried as far as desired from the generating
plant, and is therefore distributive in its tendency. As
a consequence a new social grouping seems about to
displace even the village. The telephone is coming
into use to form units of scattered farmhouses ; breaking up their isolation and allowing constant intercourse
over a wide territory. These groups are in turn linked
together ; and each to a store, a post-office, a depot,
and a physician. The tendency is steadily to increase
the relative pleasures of country life, and the advantages
of agriculture. Nature thus steps in to aid in the solution of one of the more dangerous phases of the struggle
of free institutions with ignorance.

The stability of the Republic was assured at the outset largely by the recognized collateralism and independence of Church and State. While there has been a fraternal and wholesome interaction, helpful both to citizenship and religion, these two social forces have worked without friction. A free State and a free Church, occupying the same territory, is a problem solved. Another problem, similar in character and of equal importance, is unconsciously shaping itself for action. The school will soon demand a much larger measure of independence. In fact, these three, Church, School, and State, are equally collateral evolutions of the primitive Family. The Pope, Patriarch, and Teacher have come down the ages of civil development side by side. The ideal nation will ultimately be that in which every child is born into the religious life and the school life as he is into the civic life. By birthright he should be not only a State citizen, but a church member and a school pupil. This will come about when our school system, rising from State Kindergartens to State Universities, is federated in a national University at Washington. Such a completed educational system we await not only to prevent ignorant voting, but, what is of more importance, ignorance in legislating.

Rousseau analyzing democracy, made it to consist in equality. The Declaration of Independence put it as freedom and equality. A more complete reduction seems to be, freedom to be, or to become, equal. Democratic equality is the privilege of being made capable of expressing a rational opinion by ballot. That our educational system should have remained incomplete is not surprising under the condition of such enormous political expansion as the century has required ; but it is dangerous to the stability of the Republic that it remain inadequate to the work of creating enlightened citizenship.

Third.—De Tocqueville thought he saw in the rapid expansion of the American people a cause of disruption. Grund, with keener vision, saw in the great Westward tide a bond of fellowship and an increasing nationality. The result of migration was twofold ; it weakened the provincial spirit of the mother State, and it created ties from East to West across the continent. But invariably also the newer States were more fond of the Union. Ohio and Kentucky were eminently useful in exposing and hindering Burr. In 1803, and again in 1814, Northwestern children of New England were unqualifiedly loyal to the Constitution. Emerson said in 1844, '' We in the Atlantic States by position have imbibed a European culture. Luckily for us the nervous rocky West is intruding a new and continental element into the national mind. We shall yet have an American genius.''

Almost simultaneous with the first Continental Congress was James Watts's steam engine ; and in 1787, while the Constitutional Convention was in session, Fitch launched his steamboat. The Providence that evolves ideas reveals also the means for applying them. As migration moved Westward, turning limitless wildernesses into empires, it became apparent that a new political method had been conceived an arrow's flight ahead of any before attempted. A recent advocate of centralization asks, '' What would the Fathers of the Republic have done had there been but one Colony to declare its independence ? '' It probably would not have declared independence. Our Colonies did not secede ; the Congress of the United Colonies did.

The two principles that characterized the primitive

family were Expansion and Adoption. These principles, lapsed into conquest and subjugation, have been
restored by the Federal Republic to their natural character of accretion by peaceable annexation and by
naturalization. In 1630, "The Simple Cobbler of
Agawam" lamented among other things that foreigners were allowed to enter the country, "and crowd our
native subjects." But we have welcomed twenty millions of immigrants, and assimilated them into the
population, during the last half century ;—with safety,
mainly, because of their diversity of sentiments.
America is strong across the continent because the discordant opinions of rival sections rarely reach the edge
of the country with a ripple. The votes that would
develop mischief in one direction are balanced by
others of diverse purposing.

The spasmodic efforts to restrict immigration have
fortunately been of temporary character, and limited
effect. Humanity, like all other elements, can remain
wholesome only when in rapid motion. Stability of organization depends upon instabilty of the units composing the organism. There is constant need that a body
politic, like an individual, shall "Look outward and
not inward, and lend a hand." The opposite policy
leaves the people, composing the State, to study their
troubles and brood over their ills, until these are so
exaggerated by imagination as to create the social disease of distrust. This once established, may transform
conditions of essential prosperity into a state of practical misery. It is not improbable that, when our opportunity to grow within our own territory ceases, we
shall be compelled to expand more rapidly externally.
The children of Jefferson, who purchased half a continent, and who wrote in 1823, "I have ever looked

27

on Cuba as the most interesting addition which could ever be made to our system of States,'' may be expected to have a share of his trust that the Republic is the refuge of the oppressed; and that its edge is not necessarily the waves of the Atlantic and the Pacific.

Fourth.—The present condition of the Labor Problem naturally alarms those who have not yet learned the lesson of confidence in the people rather than in leaders. At the close of the Civil War we found ourselves rapidly passing from an era of sectional conflict to an era of class conflict. We have not only legislated at the demand of labor, but equally vital legislation has subverted the rights of labor. State socialism has undertaken to regulate hours of toil, prices, and markets; while the Courts have rendered decisions believed to be fatal to justice and social equality. Labor organizations became a necessity to balance the concentration of wealth. One man is able, through capital, to outwork, and to outvote one thousand, or even ten thousand men. He may reach his hand over to the Church, and to the School, to prevent the teaching of doctrines that question his position.

Labor riots began about 1870; they have repeatedly shown what may suddenly occur in the way of antagonism to law and order. It must, however, be allowed that the tendency of labor organizations has not of late been toward anarchy but toward federation. Still the result may be, and already largely is, to displace our present mode of legislation. The rules of a Labor League, ramified through the whole United States, and formulated at a National Convention, may be as effective as an Act of Congress. Political organization is not certain to be more potent than industrial. If the latter devise laws that wisely conduct production and distribution, leading to human equality, we can with complacency view the relative decline of the power of the former. A " National Federation of Trade Labor Leagues" suggests that the Third Estate is quietly supplanting the Middle Class, as the latter displaced the Upper Class as a governing force. What shall we fear ? Our free schools have at last given us laborers who may be as well educated as capitalists. Wendell Phillips said, " Your free schools touch the bottom strata to make thinkers ; the ballot makes the same men rulers. If you do not like the consequences, you must put out your free schools."

The latest consensus of labor platforms includes, " The public ownership of all industries controlled by monopolies and trusts ; the public ownership of all railroads, telegraphs, and telephones ; all means of transportation; gas and electric plants, and all other public utilities ; the public ownership of mines, oil, and gas wells ; the reduction of hours of labor ; the undertaking of public works for the unemployed ; all useful inventions to be free to all—the inventors to be remunerated directly by the public ; the establishment of postal savings banks ; the adoption of the Initiative, the Referendum, the Imperative Mandate, with Proportional Representation, and the establishment of Co-operative Colonies." The platform of the " Federated Labor Party of Australia " demands " Electoral Reform, involving a holiday, and all public houses closed ; National Work, including State control of irrigation and village settlements ; Education, including compulsory elementary schools and optional higher schools—both free ; Industry, including statutory eight-hour day, shop inspection, and protection for exposed labor; Labor Rights, including lien for wages over work performed,

and progressive taxation of land values ; Referendum;
State ownership of railways." So we find that the
same questions agitate the whole world. Revolutionary
as many of these demands are, they are no more so
than many of the political changes of a hundred years
ago. Our Republic throbs with the half-worked-out
conviction that self-glorifying political organizations
must not prevent social organization that is capable of
considering new thought, and sifting the beneficent
from the unworkable and unwise.

Europe is ahead of us in acting on old-age pensions,
government savings banks, rescue of the gutter-born,
government control of railways, freedom of trade, referendum, and arbitration. European cities have erected
municipal tenement houses, to be leased at minimum
rents. They have taken into municipal control the
services of light, and water, and the conduct of street
railways. Municipal farms, and bath-houses, and
laundries, and milk service are being tested. France
and Belgium have not only postal banks but government pawn shops. In France thirty per cent. of the
deposits in government banks are by farm laborers ;
and in Belgium nearly half a million of children of
school age hold bank books. Three of our Postmasters-
General have urged similar banks in the United States,
for their educative force as well as their economic advantage. To lessen the extremes of poverty by pen-
sioning old age, and by State culture of the gutter-born,
seems hardly less politic. Whatever tends to create
social equilibrium forestalls danger to the Republic.

Those who enlarge on the evils involved in the
growth of the Tiers États must remember that civilization has mainly consisted in the rise of lower and
dependent classes to influential activity. The inde-

pendence of a nation is an insignificant affair compared with the rise to independence of a class. To the Middle Class of Europe, rising from servility and ignorance, we owe the best thoughts of individual and civic liberty of the last two centuries. The remaining dregs of society are now being leavened by universal free education, and by the love-faith of Jesus. Mr. Godkin, in his remarkable volume on *Social Democracy*, notes that "poor men find themselves in possession of very great power over rich communities ; and that there has grown up around this change the foreshadowing of a code of morals, in which men's right to be rich is called into question, and the spoliation of them, if done under form of law, is not an offence against morality." It would, however, be difficult to discover on American statute books any legislation, either State or National, directly originating with the poorer classes, as such, that should be looked on as an effort to despoil the wealthy. The drift to unjustly tax combines and trusts, is rather a feature of the unwise faith in our ability to moralize society by multiplicity of laws. An income tax is a part of the political system of England, and other European States ; and is meeting with general approbation. Fiat money and Free Silver did not originate with the populace. The people have only demanded a general application of the principle, laid down by the Supreme Court, that Congress has a sovereign right to make paper money the equivalent of gold. On the whole we must conclude that the present phases of the labor struggle do not involve an attack on the principles of popular government or organized society.

Fifth.—While the spoliation of the rich may become a danger in a republic, of even general education and free discussion, the spoliation of the poor exists as a

menace to which we must not blind ourselves. That
the poor are better off than formerly need not be questioned ; but this must go with the fact that the gauge of
comfort is vastly raised by free schools, and thereby the
miseries of poverty intensified. The problem is not
are the poor a shade better off, but are they better off
as educated to higher aspirations and multiplied needs.
Nor is it sufficient to show that " the ratio in our poor-
houses has declined." The number of millionaires in
the United States has increased from four in 1860 to
between four and five thousand in 1897. Meanwhile
the tramp element has gone up from none in 1860 to
about half a million. The lowest estimate, from statis-
tics collected by the *New York Journal*, covering only
part of the States, gives us three hundred and thirty-six
thousand two hundred and fifty now in the field. Mr.
Spahr, of the Surrogate Court of the State of New York,
from public records, reports that the wealthy classes
of the United States number one hundred and twenty-
five thousand families, owning wealth to the amount
of thirty-three billions ; while the poorer class numbers
five and a half million families, owning eight hundred
millions of dollars. Between these are six millions
eight hundred seventy-five thousand families, owning
thirty-one billions two hundred millions of dollars.
That is, while " Less than half the families in America
are propertyless, seven-eighths of the families hold one-
eighth of the national wealth, and one per cent. of the
families hold more than ninety-nine per cent." This
statement is accepted by Hon. Carroll D. Wright, in
the *Atlantic* for September, 1897. Mr. Shearman, in the
Forum, sums up his tables with the estimate that,
" With a generous count, far below the actual truth,
twenty-five thousand persons now possess more than

half the national wealth, both real and personal.'' Mr. Holmes, in *Political Science Quarterly*, estimates that, '' ninety-one per cent. of twelve million six hundred and ninety thousand families own twenty-nine per cent. of the wealth, while nine per cent. own seventy-one per cent.'' The *New York Tribune* estimates the list of millionaires for 1893 to be four thousand and forty-seven, '' controlling one fifth of the nation's wealth.'' Mr. Holmes adds that, '' Twenty per cent. of the wealth of the United States is owned by three-hundredths of one per cent. of the families (not including millionaires) ; fifty-one per cent. by nine per cent. of the families ; and twenty-nine per cent. by ninety-one per cent. of the families; only nine per cent. of the wealth being owned by tenant families.'' The drift to tenant life meanwhile is increasing in rapidity ; and it is estimated that less than half a century will give us an agricultural population practically devoid of freeholders. The drift in our cities is not less startling.

The solution for this concentration of wealth cannot be that given by Mr. Carnegie, who holds that some are born capitalists, and destined by Providence to control vast wealth for the public welfare. For while a large percentage of wealth is thus either used for the public, or is re-distributed at death, we see another and a larger proportion used for selfish ends, and to the detriment both of the possessors and the people. Underneath this distorted distribution lie fundamental errors of political economy and legislation. That the relations of capital and labor will be recast seems inevitable. We must at least recognize the right of labor to permanency of interest in its productive energy. The laborer now is exactly like his shovel ; to be used, and then put aside till needed again. If enterprises pay

interest, the laborer is secure in his wages; if not, he
is cut loose to find other employment, if he can.   Labor
in turn feels no obligation to the employer beyond that
which is involved in wages.   Capital assumes the entire
right to break up an industry where hundreds have
been gathered for homes and food, while the helpless
families sent adrift are not recognized by law as having
any remedy.   Senator Palmer of Illinois, while surely
not defending riot, said during the Homestead Strike:
" I maintain, according to the law of the land, not as
the law is generally understood, but according to the
principles of the law which must hereafter be applied
to the solution of these troubles, that these men had
the right to be there.   That makes it necessary for me
to assert that these men had a right to employment
there.   They had earned the right to live there ; and
these large manufacturing establishments—there is no
other road out of this question—must hereafter be
understood to be public establishments ; and the owners
of these properties must hereafter be regarded as holding their property subject to the correlative rights of
those without whose services the property would be
utterly valueless.   That concession which I make only
yields to them reasonable profits on the capital invested
in their enterprise."

The State Secretary of the Illinois Home Missionary
Society, reports of the men engaged in the coal strike
of 1897, " The men had not over eight days' work a
month, and some made less than one dollar a day, on
one third time, about eight dollars a month.   Store-
pay was going on in spite of law."   He adds that he
found " one village where the men are paid off in beer,
after the store account is settled.   In some villages
there is no school.   When the average of the year is

taken the wages are too small for the decencies of American life." Rev. D. E. Williamson adds, " The miner must furnish the oil for his lamp, powder for his blast, and ten cents every time his pick is sharpened— altogether this costs twelve cents for every ton he mines, at forty-seven cents a ton. Most of their houses have but two rooms, for which they pay five dollars a month rental." Such facts constitute a demand for a Bill of Rights on the part of the laboring classes of America quite as strong as the claim made by the people against the English government in 1688.

The danger of strikes is great, but the danger of their violent suppression is greater. This is the problem for those who imagine that the conflict is settled when hungry crowds are shot down in the streets, because they do not keep within the bounds of the law. Landon says incisively, in his *Constitutional History*, " Government may have a standing army to put down mobs ; but can popular government stand the strain of doing it ? " Mr. Watterson, one of our ablest American journalists, says, " I am afraid that organized wealth and power have not yet grown wise enough to scent the danger that is upon them. In the concentration of wealth, and in the ostentatious display of wealth, in the gradual cultivation of caste, let the wealthy behold a danger it will be well to consider in the light of both ancient and modern history." Meanwhile swarming of the discontented is growing less possible, since our government has given away the public lands ; largely to create the least manageable corporations.

But the chief danger to the United States lies in the creation of a selfish business instinct, that will supplant the heroic and altruistic in our national temperament ; and so make liberty and human progress of little con-

cern to us. It is certainly a fact that since the Civil
War our history has shown a lamentable lack of heroic
sentiment, and a dominance of greed in politics and
in social affairs. "We are suffering," says Mahan,
"from a worship of small men and small ideas." The
republic was possible only from the early cultivation of
exactly the opposite popular sentiment.

It is a mistake also to suppose that we are most in
danger of anarchy from the poorer classes. On the
contrary, combines of wealth are the most inclined, not
only to make law for themselves, but to override law,
and to an anarchical disregard for both social custom
and legal statute. Mr. Wayne McVeagh says "The
black flag of the corruptionist is far more to be feared
than the red flag of the anarchist." Organizations for
plunder do not work solely to secure favorable legislation ; they have seized upon public revenue. The
Whiskey Ring, reaching from the infinite saloons of
the land, by way of gaugers and collectors, bribed a
powerful metropolitan press, held senators in employ,
and plotted to occupy the White House. The collapse
was sufficiently sudden and awful to warn all future
conspirators. While State prisons held some of the
more notable, two men in high position committed
suicide, while others fled the country. There may be
small chance for such influences to dissever the Union ;
they might create a dry rot of free institutions, and
destroy a taste for freedom.

Sixth.—No greater danger to the permanency of the
Union exists than in the fact that government patronage of local interests has become ingrained with our
political economy. This dangerous policy entered
upon, will be escaped with the greatest difficulty, if at
all. Mr. Bryce says : "We are constantly being re-

quested, in England, not indeed by manufacturers, but by agriculturists, to introduce protection ; and our reply is this, that if we introduce it for one thing we must introduce it for others, and if we introduce it for other things, it will gain such mastery over us that we cannot get rid of it. This seems to me to be the strongest argument in discouraging any country, which is formulating its fiscal policy, from embarking in the policy of protection.'' In the United States we have so covered the question of enormous indirect taxation with the deceptive title of American Protection, that our voters are slow to perceive that it is not American to follow in the track of mediæval despotism. It is over-looked that our war for Independence was fought be-cause we were denied freedom of trade ; and that the Constitution was made to forbid impost taxes between the States. The United States presents the example of a great free-trade nation—the first free-trade nation as such—holding the banner of privileged industries.

The clash of agriculture and manufactures in 1832 brought us to the verge of scission. The reduction of tariff, brought about by Mr. Clay's Compromise, not only restored harmony but established a balance of in-dustries. In 1855 the output of agriculture was ten per cent. ahead of that of manufactures. The era of high tariff that set in with the Civil War broke up this equilibrium. In 1895 agricultural products were forty per cent. below manufactures, showing a relative change of fifty per cent. In 1855 the tonnage of our commercial marine was five millions three hundred thousand, that of England five millions seven hundred thousand. Together we ruled the ocean ; and '' the ocean always rules the land.'' In 1895 we had less than half the commercial tonnage that we had when the Constitution was adopted. Meanwhile the capital invested in manufacturing rose from half a billion dollars in 1855 to over six billions in 1895. Here we have the enormous expansion of one branch of industry, the depression of another, and the obliteration of a third. The discovery has been recently announced by Mr. Dingley that, "The real trouble in the United States is a lack of buying capacity." Back of that is the unbalancing of American industries—an evil that must be remedied before the nation can have permanent prosperity.

Agriculture not only suffers directly but indirectly ; for while it furnishes about seventy per cent. of all exports, it must reach foreign markets almost wholly in foreign bottoms. While paying a consumer's tax on imports, the tiller of the soil also pays an export tax to foreigners to get his products into the worlds markets. Even at this disadvantage his seventy per cent. of exports is the main factor in giving the country its trade balance—a balance which in 1896 was over three hundred millions.

Agriculture is still farther depressed relatively by the fact that protection breeds a system of international reprisals; and these fall not on manufactured articles, to which the tariff guarantees the home market, but upon farm produce, which seeks the markets of the world. Germany and England discriminate against American beef and pork in retaliation for a tariff that excludes their manufactured wares from equal competition in our markets. Directly and indirectly the farmer bears the burden of this false system of commercial warfare.

" A protective tariff," says Mr. Godkin, " is a tax levied on the country to indemnify a certain number of persons for their losses in carrying on certain kinds of business." Mr. Allison in the Senate asked " Is it wise to turn over the business of sugar-refining to other countries ? If not, we must tax ourselves the difference of cost in manufacturing." That is, protection compels us to pay someone to refine sugar inside geographical lines, rather than let the people buy it at a lower price when refined beyond those lines. In private business we should argue and act exactly the other way.

Never satisfying the favored classes, tariff discussion unsettles prices and leaves business to work on doubtful estimates. This fosters speculative gambling on the part of the bold, and hoarding on the part of the timid. Our history swings back and forth between periods of feverish speculation and irrational distrust. The *Globe-Democrat* of St. Louis says, " Tariff changes, even in the right direction, disturb trade ; they cause uncertainty, which is worse than the worst sort of tariff. " Giving a bonus to speculative industries has invariably led to overproduction on those lines. A glut following, has closed half the mills inside a few months, and not only dislocated commerce but filled the land with the homeless unemployed. Trusts have become necessary to prevent artificially created industries from completely breaking up in the rush of competition. These have roused the people to enact laws for unwarranted restraint on liberty.

Aiming at weakening our neighbor, protection weakens not only his producing power but his buying power; and so proves the folly of selfishness in national policy, as it is universally acknowledged to be folly in individual policy. Robbery of a neighbor's industry is a survival from the age when national morals allowed the invasion and seizure of territory.

Meanwhile, once established, protection has created lobbies which work sagaciously to increase government patronage. It has debauched representative government ; and by insidious bribery weakened the passion of patriotism. It has created a public-crib sentiment, that has turned our national Capitol into a bazaar of beggars. Civic duty has thereby dropped into the obligation to get into Congress men pledged to the protective system ; while opposition to the system is held to be un-American and incendiary. Confusing the questions of revenue and protection, the expenses of our government have been run up from a little over sixty millions in 1860 to over four hundred millions in 1896–97. At the same rate of increase we shall pay in 1930 one billion tax—if we can. Is the effort to secure a share of public plunder a nobler or a safer sentiment than a readiness to nullify statutes that do not favor local or personal interests ? The whole danger of class legislation cannot be immediately perceptible. The fierce heat that has found vent at party conventions may some day create a conflagration.

Seventh.—As soon as the Constitution was adopted the nation fell into two parts (parties) ; one emphasizing Government and Nationalism, the other Liberty and State individualism. The continued existence of two great parties means no more than that the two ideas, Individualism and Nationalism, are integral to all political life. They are to-day seen as centralism and anticentralism. Harriet Martineau wrote profoundly, " It is as inevitable that there will be always some who will fear the too great strength of State governments, as that there will be many who have the same fear about the general government. A pure despotism works indisputably. The government of the United States is disputed at every step of its working, but the bulk of the people declare that it works well.'' Between 1800 and 1860 State rights fairly balanced Nationalism. It was to the interest of both sections to assert the power of States to hold in check some Department of the General Government. Since the Civil War the drift has been much more strongly toward Washington. In 1864 Emerson said : '' I sincerely hope that this war will not prostrate the Southern States, nor greatly impair their influence in the Republic. After slavery is destroyed we shall need the South in many ways ; more especially to uphold the doctrine of State Rights, which I fear will come out of the strife sorely enfeebled.''

The position taken by the Governor of Illinois, and later by other governors, in resisting the use of United States troops in State broils, before they are summoned according to the Constitution, is the expression of a conviction that disunion, as a national danger, is succeeded by a danger quite as serious in its effect upon Federal liberty. President Grant employed United States troops in the Southern States after the war almost as freely as during the war. Fortunately he was followed by a wiser statesman, who withdrew the troops, and left each State to settle its domestic affairs as best it might. President Grant's Federal interference was as futile in good results as President Hayes's policy was speedily successful in restoring order. The first duty of a Governor certainly is to comprehend the rights of his State, and to uphold them. Our national government is not only Constitutionally limited in its powers of interference, but its efficiency depends on not exceeding those powers.

The problem of reconstruction following the Civil War was the most complex that ever insisted on imme-

diate solution. Had the Southern States really succeeded in seceding, for a time, from the Union? Or,
as Brownson shrewdly urged, had State-suicide occurred? It is notable that several writers have recently
spoken of " the Union as it was," and " the Union as
it is." Did reconstruction really introduce any new
principle? Was the whole Union reconstructed? Is
there now left only a single nation, composed of States
with less State Rights than they formerly had? Has
the Central Government an increase of rightful powers?
If so, which Department is magnified? Certainly no
amendment of the Constitution has decreased the
powers of the States, except so far as concerns the
possibility of holding slaves, of restricting suffrage, and
of denying equal protection of the laws.

Still we cannot avoid perceiving that there has been
a very great change in the relation of the General
Government to the States. Force Bills, to supervise
State elections, were enacted as late as 1874. In 1888
the scheme was renewed, by the action of the Senate ;
and not till the administration of Mr. Cleveland, in
1894, was a finality put to this sort of sectional arrogance. So sharply did our political pendulum swing
from disunion to consolidation.

Much that was done, and defended in its doing, as
war power, remains in the body politic. John Quincy
Adams first advanced the idea that while the General
Government had no right to meddle with slavery in
time of peace, in war the case would be different. Lincoln acted on this principle when he said, " What I do
about slavery, I do because I believe it helps to save
the Union." " But," says Judge Cooley, " the act of
Mr. Lincoln went a long way toward creating a general
impression that expediency is legality." The Supreme

Court had been brushed aside ; and Congress was not consulted as to the abolition of the chief social institution of a third of the States.

For the purpose of increasing the revenue, Congress created a national banking system, and taxed to death State banks to make room for it. It issued money, which the Supreme Court deliberately pronounced that Congress, by right of sovereignty, could make the equivalent of gold ; and could compel creditors to take for debts contracted on the coin basis.

The power of the Supreme Court, while nullified at will by the President and by Congress, was greatly extended over the State Courts. This power it holds and exercises to the present day, to its own engorgement, and to the detriment of national business, as well as the derogation of rightful State sovereignty. For increased revenue the tariff was raised to a basis that was then held excusable only as a war necessity. The result was not only enlarged revenue, but the enrichment of a horde of speculators, who have since constituted a peculiar factor in our social organism. Once raised to an extraordinary height, indirect taxation was not so easily reduced. Since the war the tariff has been twenty-four times changed, up to 1897 ; and in all cases but one, with more or less increase of its power for taxation.

The duty of providing for disabled soldiers opened a door for extravagant outlays, under the cover of national gratitude. The pension list has therefore grown rather than diminished, for over one third of a century. The roll now stands 983,528, including an increase of 12,850 for the fiscal year ending June 30, 1897. The evil has been both the pauperization of public sentiment and a growing habit of lavish public expenditure. National expenses are nearly sixfold what they were in

1860, while population has only a little more than doubled. We pay taxes more readily, and we eat at the cost of the public with less shame. Thousands, who would scorn the public poor-houses, eat the public dole in their own cottages—or even mansions. The people have begun to speak of the government as something apart from themselves ; its obligations and duties being no longer identified with their own.

At the close of the nineteenth century we are essentially discussing the very problem that closed the eighteenth. It is the old struggle of Hamilton and Jefferson over again. The present volume may have thrown some light on which of these men was the wiser leader. A study of Madison, on page 100, will satisfy us that the intention of the founders of the Republic was, A Federal Union of Independent Sovereign States. This granted, we must consider whether the drift toward centralization must be accepted, on the ground that the Constitution is, as some writers suggest, '' outgrown.'' Andrews, in his *History of the United States*, has not hesitated to say that, '' While men still differ as to the original nature of the Union, yet the war laid the question of National supremacy over State forever at rest ;' having therefore virtually the effect of a Constitutional amendment. Practically the war entailed enormous new exaltation and centralization of the Union, with answering degradation of the States.'' In other words war power has amended the Constitution, and altered the character of the Union. Are we a military Republic ? A growing tendency to criticise the national Senate, and demand its reconstruction on a popular basis, tends in the same direction of consolidation. Dr. Von Holst urges the unwisdom of allowing '' the legislature of Nevada, with a popu-

lation barely sufficient for a third-class city, to delegate two men to stagnate seventy millions." General
Trumbull urges the absurdity of allowing those same
Senators, "representing forty thousand people, to pair
with those of New York representing six millions."
But this inequality was fully as well understood in
1787, when Delaware was allowed to balance Virginia,
New Jersey to balance Pennsylvania, and Rhode Island to balance Massachusetts. The key of the Federal
Union is State equality as well as the equality of all
citizens. If the Federal principle ceases to hold as
fundamental in the Union, we have yet to discover
any other principle that will bind together a whole
continent in political fellowship.

Eighth.—The capacity of the people for self-government altogether has of late been challenged by as able
disputants as Sir Henry Maine and W. H. Lecky.
Contrasts are drawn between the government of England under Victoria and that of the United States for
the same period of time, with confidence that the latter
has not proved to be either as progressive or as free
from corruption as the former. Herbert Spencer an-
nounces that he has lost much of his former faith in the
political intelligence of the people, holding that it will
be a long time before they will be capable of the wise
conduct of the State. Even Lieber doubts the doctrine
of *vox populi vox dei*. Bluntschli asserts that " the
populace cannot retain virility after having drunk the
intoxicating wine of power." Do the people come
under the law of what Freytag denominates " The
Disease of Monarchs"—that is moral imbecility?
I believe that a fair review of the century will
justify John Adams when he said, " All governments
are of one, of a few, or of all ; and we believe the

latter, although attended with evils, the safest." Gallatin said, " I will rest on the people as a full security against every endeavor to destroy our Union." The people have stood the test invariably better than their leaders. "American representatives," say Mr. Goss and Mr. Lowell, " have retained more purity, and are less purchasable, than American Senators." " You can trust honest hearts with weak heads better than wise heads with dishonest hearts." Still the habit of congratulating ourselves on having safely passed an exciting political campaign indicates at least a consciousness of danger. Probably the century may be fairly summed up in the words of Disraeli :

" The People, Sir ! are not always right ! "

" The People ! Mr. Gray, are seldom wrong."

Reviewing our legislative assemblies, Mr. Godkin believes " there has been a decline of the quality of the members in general respectability, in education, in social position, in morals, in public spirit, in care and deliberation, and I think in integrity also." But nearly one hundred years ago a member of the first Congress after the Constitution said in a letter to another, " What a damned set of rascals we had in that Congress." Yet those Congressmen were largely the Federalist's " Best." In 1796 Oliver Wolcott, of Washington's Cabinet, wrote, " I believe there never was a public body that deserved less of public confidence—who were more ignorant, vain and incompetent, than the mass of the present House of Representatives." Rings and boss rule have not increased in power, and certainly not in malign influence, since the days of the Patroons, of Burr, and Hamilton, and George Clinton, down through the rule of Van Buren and Thurlow Weed to Roscoe Conkling.

A greater danger consists in the growth of the passion for law-making. Our codes overflow, and no one knows their full contents. Every legislator is estimated by his ability at formulating statutes and " putting them through." In this way business rests on uncertainties, while no remedy is nearer than counter-legislation, itself liable to be soon reversed. This excess of legislation has carried us over from the tyrannical inflexibility of tradition and precedent to the equally tyrannical instability of fresh laws every morning. Our system aggravates the difficulty by allowing a change of opinion in a small percentage of voters to change the legislative majority and the total character of legislation. No Article of the Constitution restrains Congress from passing over, at a single sitting, from a system of Free Trade to a system of High Tariff. We are now living a generation behind the best advantages of the age because a powerful lobby procures legislation that withholds the most important inventions from free use by the people.

It is believed by many that Proportional Representation, with Referendum, would largely check the superfluity of legislation. It seems more probable that we shall be compelled, more and more, to turn to Legislation by Experts. Educational affairs should, it would seem, be regulated by experienced educators—as affairs of the church are given over mainly to church legislation. Mr. Godkin believes we shall before many years " get our government more largely from constitutional conventions, and confine legislation within narrow limits." We are already moving on the lines of restricted sessions ; and State Commissions to regulate specific interests.

The Executive Department of government, which

was most feared by the framers of the Constitution, has proved least dangerous to liberty. This has been due, first of all, to the ideal of patriotism established by Washington, and sustained by his successors. Their example has stood as strong and as influential as the Constitution. The brevity of the term of office has proved an additional safeguard ; while the counter-check of the other Departments has worked largely as was anticipated. That the Executive Department may, however, even unwittingly, lead to disaster, we have only to recall the unfortunate Venezuela Message of Mr. Cleveland. Reading backward the noblest event of the century, the Monroe Doctrine was made to en-danger the peace of the Anglo-Saxon world, and to saddle the United States with a Protectorate over two continents. Coincident with this accident, was the offi-cial correspondence, carried on by the same President, with the dethroned ruler of Hawaii—after the Repub-lic of Hawaii had been acknowledged by the United States and leading European Powers—correspondence that must be classed as international treason. The probability, however, of such Executive action lead-ing the Republic into vital errors, was answered by the almost universal protest of the American people, leading to a broader affirmation of the principle of Arbitration.

The trend of previous chapters has been so strongly to emphasize a danger to the perpetuity of republican government from the exercise of arbitrary power by the courts, that I shall recall the subject here only to quote from others.

Justice Harlan, while handing down a dissenting opinion on the Income Tax, said : " It strikes at the very foundation of national authority. It may provoke a contest, which the American people would have been spared, if the Court had not overthrown its former decisions. Congress cannot tax incomes, while it may compel the workman to contribute directly from his earnings for the support of the government." Justice Brown said, " I cannot escape the conviction that the decision is fraught with immeasurable danger to the future of the country, and approaches a national calamity." Justice White more vigorously still, said that, " If such a system were followed, the red specter of revolution would shake our institutions to their foundations."

David McG. Means, discussing the question, " Will Government by the People Endure," says: " Of all the checks upon misgovernment, the Supreme Court has been regarded as the strongest and surest ; and it is still spoken of as the palladium of our liberties. But from the evil day of the first legal-tender decision, thoughtful men have seen that its foundation had been undermined. The Constitution and membership of the Board were altered by Congress, and the President; if not with the deliberate purpose, at least with the foreseen result, of procuring a reversal of judgment, on perhaps the greatest Constitutional question that has ever come up for decision. It is hardly speaking too strongly to say that this proceeding changed the nature of our government. Not only was the issue of the fiat money of the Civil War declared Constitutional, but, as a sequel, our government was declared to be, not of limited powers, but to possess the absolute authority of the despotisms of other continents. Not only in time of war, but also in peace, Congress has now plenary power to substitute irredeemable paper for silver and gold ; and to compel every citizen to accept the substitute in payment of all debts."

The same writer adds, " Unless some check can be put upon the abuse of government, peril will occur. If the Constitutional party insists on the issue of money by the government, the Radical party will demand the same rights. If laws are passed for the profit of the intelligent and wealthy classes, the poor and the ignorant will demand laws in their favor. If Congress can impair the obligation of contract, by making government paper a legal tender, it can certainly make silver a legal tender. Pensions, protective tariffs, silver bounties and greenbacks may seem desirable to respectable citizens, so long as their party is in power ; are they prepared to have the principle carried out by their opponents ? "

Added danger arises from the sentiment, that has found strong lodgment, that while to criticise the other Departments of Government is the most positive duty of a citizen, it is a civic crime to criticise the Judiciary. Judge Cowing of New York, in 1896, said, " He who assails the Judiciary becomes the disturber of public peace and order, and is an enemy of the government. Such a man should be regarded as a pirate." This is a revival of the spirit of the Alien and Sedition Laws of one hundred years ago. Charles Sumner, in a similar case, responded, " I hold judges, especially the Supreme Court, in much respect, but I am too familiar with the history of judicial proceedings, to regard them with superstitious reverence."

Reviewing the first century of national life, we must be convinced that it is rather the Declaration of Independence than the Constitution that has outgrown the limit of its first expression. We have added to it the *right of all men to be educated ;* and are now trying to formulate the additional *right of all men to labor* if they will, and to share equitably in the result of their labor. That this great new principle will shape itself into our national life is beyond question. "Work for the unemployed is the first call to duty, and demands immediate action. To rescue these from tenements and hovels, from streets and slums, from charities degrading bondage, and give them the opportunity of applying their labor to the natural resources, is the initial and commanding duty of the present hour." The American nation will be compelled to thoroughly consider the social law stated by Cardinal Manning, "The obligation to feed the hungry springs from the natural right of every man to life, and to the food necessary to the sustenance of life. So strict is this natural law that it prevails over all positive laws of property." The social ethics of the Twentieth Century will certainly be nearer that of Jesus Christ.

While optimism and patriotism must not blind us to the danger that threatens the perpetuity of Popular Government and the Federal Union, there is no reason for believing that we are not still sound at the core, and abundantly able to adjust ourselves to social progress. With more flexibility than a monarchy, the Federal Republic has so far been able to invite discussion and endure agitation. Our great need, as we close the century, is a revival of heroic sentiment, and a leader able to measure the new Republic—which spans the continent, and begins to weigh the islands in the ocean. Will another Jefferson arise for 1900 ?

APPENDIX TO CONCLUDING CHAPTER

JUDGE T. M. COOLEY ON CENTRALIZATION

During the Civil War many clear infractions of the Constitution were excused by the public as being justified by an overruling necessity ; such, for example, as the interference by federal forces with State elections in Kentucky. The longer the war continued, and the more numerous were the excesses of power, the more they came to seem in the minds of many persons to be in harmony with the spirit of a Constitution, which was designed to insure the perpetuity of the Union, and might therefore be supposed to contemplate the doing of whatever was essential to that end. " We break the Constitution that we may save it," was sometimes said ; a paradox, the mischief of which was not universally perceived until calmer times brought cooler heads. It was the opposite view—that the Constitution might be appealed to for protection even by those who were seeking to destroy it—that seemed at the time preposterous. When, therefore, men were tried and condemned for treasonable practices before military tribunals in Indiana, the proceedings were approved by a prevailing contemporary sentiment, which held that the protections to liberty incorporated in the Constitution were subject to an implied exception, and might rightfully be set aside when great emergencies required it.

Many such things are inseparable from a state of civil war ; and they are recorded afterwards not so much for the purpose of fixing the responsibility for them upon individuals, as to guard against their being accepted as

lawful, and thereby leading to mischief in the future.
But in this connection they are to be noted also as ac-
counting in some degree for the rapid strengthening of
federal power while the war was in progress. A violation of the Constitution, even when disapproved by
public sentiment, may nevertheless have important influence upon the public mind, in accustoming it to
accept as quite in order other questionable acts which
before would have been promptly condemned. A
wholly baseless claim vigorously insisted upon, especially when the power of present enforcement exists,
may be as likely in public affairs as in private business
to lead to compromise by concession of some part of
what is claimed.

But the centralizing forces which raised no question
of constitutional right or authority were now powerful.
The government was making vast military expenditures ; it was giving out enormous contracts in which
the profits might be large, and the birds of ill omen
gathered about the departments in great flocks, as
eager for their feasts and as reckless of anything else
as the vultures upon the fields of battle. The government was all the while drawing in and paying out large
sums of money ; and the financial currents were to and
from Washington, not to and from the State capitals,
except as the States were acting as subordinate auxili-
aries in the war. Many new offices were now necessarily created ; and for the time being the National
Government was the great dispenser of favors, privileges, valuable employments, and profitable contracts,
whose executive, by a dash of the pen, was giving
offices which gratified the ambition of a lifetime, while
heads of departments by their favors were enabling
others to lay the foundation of enormous fortunes. All

these things not only for the time affected the relative
interests of the people in their State and National Governments, but they greatly and permanently affected
the imaginations of the people ; diminishing the States,
and their rights and powers relatively to the Union,
and making them appear in a constitutional point of
view, less and less like sovereignties and more and
more like subordinate sections of a State.

Nothing in this regard affected the imaginations of
the people more than the destruction of the institution
of slavery in a considerable portion of the Union by
Executive Proclamation enforced by the army. The
question in Mr. Lincoln's mind, how slavery should be
dealt with, had become one of mere expediency ; and
when he decided, as he shortly did, that the destruction of slavery would conduce to the restoration of the
Union, he gave the fatal blow. It may have been an
act of questionable constitutional right, but it was irre-
versible when done, and it went a long way in strengthening the growing impression that in time of war
whatever in government is found expedient must be
legally admissible.

Then Congress undertook—what it had never attempted before—to provide the whole currency of the
country. It had power by the Constitution to coin
money ; but coin had always constituted a small percentage of the whole currency, the most of which had
been the bills of State banks. Twice a National bank
had been chartered as an expedient agency in government ; but the constitutional power had always been
contested, and, though affirmed by the judicial, had
been denied at last by the political departments of the
government under the lead of Jackson ; and the judgment of the people might be said to stand recorded

against the judgment of the court. But now Congress assumed to give corporate powers not to one National bank merely, but to banks in every quarter of the country, sufficient in number for all the demands of business ; and the question of power to do so was scarcely made in any quarter. Congress did not stop at authorizing national banks ; it undertook to destroy the State banks to make place for them. It was not claimed or pretended by any one that this might be done directly and avowedly ; for State power to create banks of issue was unquestionable, and what the States had lawful power to create, Congress could not have lawful power to destroy. A very obvious comment is, that if one class of State corporations may constitutionally be thus legislated out of existence by Congress, that body must have the like power to destroy at pleasure other State corporations ; and it might, perhaps, on some view of national expediency, tax out of existence all corporations for insurance purposes, except such as Congress itself might charter for the District of Columbia and other territories and places within its exclusive jurisdiction ; thus taking to itself this whole subject as completely as if control over it had been expressly conferred. This would be making the power given to Congress for the purposes of revenue, a power of destruction irrespective of revenue. But the tendencies of the times were such that this legislation was sustained, with little question and less opposition. The feeling was general that the country was well rid of State bank bills, which in times past had been infinitely mischievous, and nobody troubled himself with the question whether a dangerous precedent was not being established in the process of getting rid of them.

The government also issued bills of its own, and declared that they should be legal tender as between individuals ; not merely for such debts as should be thereafter contracted, but for pre-existing debts, contracted when gold and silver alone were legal tender. Then came the question—Whence did the government derive the power to give this effect to the evidences of its own indebtednesss ?   It is not to be found expressly conferred by the Constitution ; there is nothing in the debates of the Convention which framed that instrument, indicating a purpose to confer it.   Legislators and lawyers looking for it in the Constitution, suggest that it may be referred to the power to borrow money, or the power to coin money, or to some other specified power ; but at any rate it may be referred to the war power, which is so tremendous in its scope that those wielding it can alone set bounds to it.   If in their opinion the issue of legal tender currency is a necessary expedient when war puts the existence of the Union in peril, then the issue must be as lawful as the employment of men or artillery in the field.   Such was the reasoning of many at the time.   But when it is once determined that the power of Congress may be grounded in necessity, it logically follows that it cannot be limited to the time of war.   The necessity that makes for itself the law, knows no times ; it is conceivable that it may be slight in time of war and urgent in time of peace ; and when the groundwork of right is admitted, the power which passes upon the necessity cannot be restricted in the occasions.   And necessity under such circumstances can mean only expediency.   We thus reach a stage when Congress, on its own view of expediency, may exercise the tremendous power over contracts, of making them payable in something besides the money which the parties understood they were bargaining for : something which may or may not be of equal value ; though if it were of equivalent value, there could in general be no occasion for imparting to it the legal tender quality.

The war made heavy taxes a necessity ; and the government, following its ordinary course, raised these for the most part as indirect taxes. In so far as they were levied upon imports, the levy afforded opportunity to discriminate for the protection and encouragement of American products. The heavy tariff thus became in large degree a protective tariff. When the war was over, a fearful load of national debt remained, and the war taxes were continued for the gradual extinguishment of this debt. But when the debt had so far diminished that the heavy taxes could no longer be defended on that ground, the protected industries were found to be so numerous and so powerful, that they were quite able to prevent success in any attempt at considerable reduction. The tariff thus became distinctively a tariff for protection ; and all the protected interests looked to the Federal Government as being at once, to some extent at least, the source and protector of their prosperity.

The superabundant revenue that has come to the government as a result of heavy taxation, has made Congress over-liberal in the matter of expenditure. Schemes of doubtful public utility have easily found support ; proper national works have readily obtained extravagant appropriations ; and sometimes it has seemed that money was voted without discrimination, so many of the persons who cast the votes appearing to look for the benefit in the tax from which the money came, rather than in the purpose which was to be accomplished by its expenditure. To every locality that has received a grant from the General Government, the grant has somehow seemed like a mere gift, as if in some providential way the money had come to the national treasury without cost to the people, and the nation was distributing it in benefactions. The State would be powerless to make such benefactions except at a cost of direct taxes ; and the people of the State would never assent to the levy of taxes for such purposes. In fact, they have prohibited it by their Constitution.

An overflowing national treasury has also encouraged liberal pensions, and gradual additions to the classes of pensioners, until the number dependent upon the nation for bounty of this nature has become enormous.

The nation has also since the war made gifts of vast areas of land for the construction of railroads, and loaned large sums of money which might also as well have been made gifts. It has added to its postal service something of an express business, which has within it the prophecy of greater things to come. The question of annexing the telegraph to the postal service is being urged, and the question of the nation assuming the regulation of railways has for some time been before Congress, and is certain to receive at some time in the near future an affirmative solution.

The number of Federal office-holders has increased until they constitute a mighty army ; an army greater in number than that with which Wellington at Waterloo changed the history of the world ; greater than that with which Meade won the decisive victory at Gettysburg in the crisis of the Civil War. It has been deemed necessary to legislate to prevent elections from being improperly influenced by the labors and pecuniary contributions of so large a body, directed and expanded as they are likely to be, by the political machinery of the party in power.

Everything gravitates to Washington ; the highest interests and the most absorbing ambitions look to the national capital for gratification ; and it is no longer the State but the Nation that in men's minds and imaginations is an ever present sovereignty. And this is as true of the States of which Jefferson and Calhoun have been the idols as it is of Massachusetts or Michigan. —From *History of Michigan*, published by Houghton, Mifflin & Co.

FROM ADDRESS ON INDIVIDUAL FREEDOM, BY THOMAS
   F. BAYARD, BEFORE THE EDINBURGH PHILO-
   SOPHICAL INSTITUTION

I have witnessed the insatiable growth of that form of State socialism styled " Protection," which I believe has done more to foster class legislation and create inequality of fortune, to corrupt public life, to banish men of independent mind and character from the public councils, to lower the tone of national representation, blunt public conscience, create false standards in the popular mind, to familiarize it with reliance upon State aid, divorce ethics from politics, and place politics upon the low level of a mercenary scramble, than any other single cause.

Step by step, and largely owing to the confusion of civil strife, it has succeeded in obtaining control of the sovereign power of taxation; never hesitating at any alliance, or the resort to any combination that promises to assist its purpose of perverting public taxation from its only true justification and function of creating revenue for the support of a government of the whole people, into an engine for the selfish and private profits of allied beneficiaries and combinations called "Trusts."

Under its dictation individual enterprise and independence have been oppressed, and the energy of discovery and invention debilitated and discouraged. It has unhesitatingly allied itself with every policy which tends to commercial isolation, dangerously depletes the Treasury, and saps the popular conscience by a scheme of corrupting favor and largesse to special classes, whose support is thereby attracted.

More than seventy years ago when this practice of the substitution of State interference for free individual enterprise was first mooted, and before the destructive policy of protection had struck its root in American legislation, Daniel Webster had said, "How, sir, do ship owners and navigators accomplish this? How is it they are able to meet and in some measure overcome universal competiton? Not, sir, by protection and bounties, but by unwearied exertion, by extreme economy, by unshaken perseverance, by that manly and resolute spirit which relies on itself to protect itself."

Once our laws encouraged enterprise on land and sea, and our people succeeded. The four pillars of our prosperity were agriculture, manufactures, commerce, and navigation. American ships and American trade went everywhere ; and the American flag at sea had the respect of the world. So it was, but is no more. No call to the sea sounds now for young America ; it has been swept by rivals so cleanly from the sea that the proportion of American carriage in what is called American commerce is now but half as large as in 1789.

When we had shipping of our own and merchants

of our own people to carry on our trade we had no fear of adverse balance and the export of gold. There was then no nation, rival, or enemy, that could strip us of our wealth. Now there is, and the world knows that a foreign marine is a stripping machine. A famous Englishman laid down this maxim for his country's guidance : "Whosoever commands the sea, commands the trade ; whosoever commands the trade of the world commands the riches of the world and, consequently, the world itself."

It is incorrect to speak of " Protection " as a national policy, for that it can never be ; because it can never be other than the fostering of special interests at the expense of the rest ; and this overthrows the great principle of equality before the law, and that resultant sense of justice and equity in the administration of sovereign powers which is the true cause of domestic tranquillity and human contentment. The value of " protective " legislation to its beneficiaries consists in its inequality ; for without discrimination in favor of someone there is no advantage to anyone, and if the tax is equally laid on all, all will be kept upon the relative level from which they started ; and this simply means a high scale of living to all, high cost of production of everything, and consequent inability to compete anywhere outside the orbit of such restrictive laws.

But the enfeeblement of individual energies and the impairment of manly self-reliance are necessarily involved; and the belief in mysterious powers of the State and a reliance upon them to take the place of individual exertion, fosters the growth of State Socialism, and personal liberty ceases to be the great end of government.

INDEX